SYSTEM SAFETY
ENGINEERING AND
MANAGEMENT

SYSTEM SAFETY ENGINEERING AND MANAGEMENT

SECOND EDITION

HAROLD E. ROLAND
University of Southern California,
Los Angeles

BRIAN MORIARTY
Dumfries, Virginia

A WILEY-INTERSCIENCE PUBLICATION
John Wiley & Sons, Inc.
NEW YORK / CHICHESTER / BRISBANE / TORONTO / SINGAPORE

In recognition of the importance of preserving what has
been written, it is a policy of John Wiley & Sons, Inc. to
have books of enduring value published in the United
States printed on acid-free paper, and we exert our best
efforts to that end.

Library of Congress Cataloging in Publication Data:
Roland, Harold E., 1942–.
    System safety engineering and management / Harold E. Roland, Brian
Moriarity. — 2nd ed.
        p.   cm.
    "A Wiley-Interscience publication."
    Includes bibliographical references.
    ISBN 0-471-61816-0
    1. System safety.   I. Moriarty, Brian.   II. Title.
TA169.7.R64   1990
620'.001'1—dc20   90-12109
                                                                  CIP

Printed in the United States of America

10 9 8 7 6 5 4 3 2

PREFACE

This book is intended to be the practicing system safety professional's reference manual. It describes the field of system safety as it is presently conceived and practiced, that is, imposing a state safety on systems under design. System safety includes within its purview, evaluating a system for its state of safety over its life cycle, determining if that state is acceptable, evaluating countermeasures for their effectiveness in bringing the system to an acceptable state, and making decisions concerning the employment of a set of countermeasures and the use of the system. It thus contains elements of the field of risk assessment and management, except for insurance or risk transfer aspects.

These methods are also useful in fully operational systems so this book will be helpful to those safety professionals who are examining systems completed and deployed. The traditional industrial safety professional who is concerned with protection of the worker in the work place, will also find these methods useful.

This book does not contain all of the elements of safety, particularly the *a posteriori* (after the fact) methods and practices. The focus in system safety is to prevent accidents before they happen, thus this form of the practice of safety may be considered *a priori* (before the fact). The form of safety that has generally been followed since safety evolved as a discipline in which a *known* precedent derived from an accident is used as a basis to prevent further accidents, is inadequate to deal with accidents for which the consequences are too catastrophic to be borne. System safety attempts to anticipate accidents and judge the total risk of the system over its life cycle. It is able to do this through a unique set of management practices and analytical methods.

This work describes both the management and analysis methods that can be directly applied to achieve a satisfactory state of safety. Although this is not a statistical text, sufficient statistical methods are described to bring the safety professional to a state of knowledge where the analytical methods can be understood and used. This updated text briefly describes other quantitative methods that may be used in safety data analysis. The authors have also included several decision and evaluation tools from engineering and business that are most useful in safety evaluation and risk analysis. This text contains the latest thinking in the fields of fault tree analysis, sneak circuit analysis, and software hazard analysis as well as recent changes in industry requirements for system safety management and documentation.

The methods and practices described attempt to strike a medium between those prescribed by government agencies such as Department of Defense, Department of Energy, and National Aeronautics and Space Administration, and general industry practices.

HAROLD E. ROLAND
BRIAN MORIARTY

Los Angeles, California
Dumfries, Virginia

CONTENTS

I

MANAGEMENT

1

INTRODUCTION

To be free of peril is a universal goal that has been common to all eras and all peoples. The desire to be safe and secure has always been an intimate part of human nature. But it is not enough to satisfy all our needs. People search continually for ways to improve their creature comforts and way of life, exploiting the environment to meet their needs. The ability to improve their surroundings increased immensely when they used the first tool, and then better techniques to fashion innovation were devised. Humanity soon became *technical* in the means to provide the objects and conditions necessary for sustenance and physical contentment. In this sense, technology may be viewed as the changing environment of humanity. The role of safety has played a vital part in this development. It has a history of its own.

The idea of safety has existed since the dawn of human history. Early people had good reason to exercise defensive caution: they feared the many natural hazards around them because they did not understand what caused them. They perceived the obvious risks of the predatory beasts and were always aware of the comparative strengths of their neighbors. Like the animals they hunted, they developed a keen innate ability to sense a threat, and learned to evaluate the dangerous and to react protectively. Without question, prehistoric people acquired the ability to devise a plan for safety, which had become a vital part of their survival.

As time went by, the safety plan took on a social form.

To satisfy their security needs, people began to adapt, then to discover, and finally to invent. All of these efforts gave them the ability to advance further and to accelerate desired changes. The advances usually had negative as well as positive effects on values, for although innovations satisfied their needs, the changes often presented unforeseen factors that menaced or damaged segments of the society. Table 1.1 lists a few examples of how technological advances can create hazards (Jerome, 1976).

Humans must deal with those hazards that occur through carelessness and the unguarded or inadequately organized uses of their devices and substances. They

3

Table 1.1. Technology versus hazard

The Technological Advance	The Hazard
Fire	Burns, conflagrations
Knives	Inadvertent trauma
Fossil fuels	Atmospheric pollution
High speed transportation systems	Accident induced property damage and injury
Pesticides	Food chain toxicity
Food preservatives	Carcinogens
Nuclear energy	Ionizing radiation

must develop controls to minimize hazards. Changes often occur in social behavior when we are faced with a threatening environment. These changes, coupled with social controls, are used to assimilate new developments into constructive functions. Safety in aviation and the space program represents one of the more sophisticated outgrowths of such assimilation.

It took centuries of difficult experience for us to perfect the discipline of safety, a discipline in which our resources must be so well organized as to achieve optimal loss over the entire life cycle of the systems we develop. It was such a safety discipline that assisted in putting men on the moon. Such disciplines made air transportation so safe that it is the common way to travel over long distances. Attaining such a technical state requires a vast cache of knowledge in both technology and management. And, more important, we now have learned methods of avoiding the pitfalls of human weakness. A modern poet put it very clearly:

> The road to wisdom? Well it's plain and simple to express: Err and err, and err again, but less, and less, and less.
>
> —Piet Hein, 1966

The concept of system safety advances one giant step from this concept. It stems from the *logic of system functions* and of *preventing errors before they happen*.

Today we find that the safety of each person at home, at work, in transit, at play—in one's entire environment—is the common responsibility of the individual industry, the community, and government. The first significant steps in safety's evolution were oriented toward industry and were taken about the time of our country's Centennial. The need for safe working conditions was brought to light at the time by changes wrought during the first decade of the Industrial Revolution, and its acceptance was assured through innovation by the pioneers of modern management. Enhancing industrial safety was found to be cost effective, as well as morally obligatory.

> It is immoral to design a product or system for mankind without recognition and evaluation of the hazards associated with that product or system
> —*Statement attributed to an anonymous early safety professional*

The past century produced knowledge that has assured each individual a reasonable degree of safety in the overall environment. More important, it was also found that safety could be cost effective for the whole of society. However, statistics published each year by the National Safety Council reveal that the United States is caught up in a repeating accident syndrome that is painful. The number of deaths and disabling injuries has reached 9 million annually and is growing. The cost of these accidents is upwards of $130 million. These figures do not include unreported accidents, indirect costs, or any of the consequences resulting from long-term, low-level exposure to hazardous substances. Thus, a principal theme of system safety is to ensure that we know what the hazards are before the system is allowed to operate. In this way, we are prepared to accept the residual risk of using important systems that benefit mankind while buffering the disruptive consequences of the occasional accident. Just as the designer has the ability to learn from his or her mistakes, mistakes in the form of accidents will occur. In the past decade, the Three Mile Island incident, asbestos manufacturing and installation, recent train collisions (Amtrak/Conrail 1987), the Chernobyl nuclear power plant disaster, and the Challenger space shuttle vehicle O-ring destruction are but five examples that are evident to us. Our concern is what could have eliminated these events. When the investigations are finished and the final papers are written, the presence of a logically based System Safety Program will become the main thrust of a prevention activity.

Technological advances present more complex methods of design and integration of hardware and software. Nevertheless, the analytical tools and resources available to us provide us with the ability to understand these systems and assure ourselves that their hazards are known.

1.1 DEFINITIONS

Recent accomplishments of the space shuttle, new and advanced weapon systems, larger and more advanced aircraft systems, nuclear power systems, and computer developments for command and control systems, have all shown the need for a better understanding of accident prevention. Inherent in the use of these systems is the recognition of hazards and their identification. With modern improvements in science and technology, there is the constant and growing need to learn about hazards and their means of control. Safety practitioners have observed the need for sound hazard identification management and a disciplined program of hazard control. The first step in this total process is *hazard identification*.

The hazard identification process requires a disciplined method of analyzing the product or system. The discipline of system engineering has been a major part of organization for systems development. This organizational development emphasizes systems methodology—that of *system safety*—which is established as a formal, disciplined approach to hazard management. We will discuss this new discipline, its concept, and its background. To ensure that we understand the basic terms for hazard control, let us discuss the definitions as applied to the methodology known as system safety.

1.1.1. The Hazard

The word *hazard* has many definitions listed in the dictionary; among them are:

1. risk; peril; jeopardy   2. a source of danger   3a. chance, b. a chance event; accident   4. *obs*: mistake   5. something risked

The safety person sees a *hazard* as an implied threat or danger, of possible harm. It is a potential condition waiting to become a loss. A stimulus is required to cause the hazard to transfer from the potential state to the loss. This stimulus could be component failure, a condition of the system (pressure, temperature, switching condition that is out of tolerance, a maintenance failure, an operator failure, or a combination of other events and conditions. The following is a more technical definition of a *hazard*:

A potential condition, or set of conditions, either internal and/or external to a system, product, facility, or operation, which, when activated transforms the hazard into a series of events that culminate in a loss (an accident). A simpler and more fundamental definition of *hazard* is a condition that can cause injury or death, damage to or loss of equipment or property, or environmental harm.

1.1.2. The System

The word *system* is worthy of some definition. *System* is used throughout this book and the literature with two meanings. The first, referring to a new design upon which a state of safety is to be imposed, is the more familiar and the primary dictionary definition.

A group of interacting, interrelated, or interdependent elements forming or regarded as forming a collective unity.

A more direct definition of *system* is a composite of people, procedures, and equipment that are integrated to perform a specific operational task or function within a specific environment. A *subsystem* represents an element of a system that may constitute a system in itself.

1.1.3. The Stimulus

The normal system has a hierarchy of *subsystems, assemblies, subassemblies,* and *components.* The number of these elements in the hierarchy will vary with the type of system and its complexity. Naturally, the components are interconnected in such a manner that they perform a specific function when input is provided from a source, such as another component or a human operator. Thus, an action of the component will create output. This series of interconnected events cause a sequential, logical action in the system design. As an example, take the action of energizing a relay, which will force closure of the relay switch as a specific performance level of the relay, dependent on its characteristics in amperage and loading requirements to the relay coils. We see the event, closure of the relay, resulting from the *stimulus* of current applied to the relay coil. We can thus define the *stimulus* as: a set of events or conditions that transforms a hazard from its potential state to one that causes harm to the system, related property, or personnel.

1.1.4. The Accident

We normally think of the accident as the loss of a system or part of a system, the injury to or fatality of the operators or personnel in near proximity, and property damage of related equipment or hardware. An accident is usually a dynamic event since it results from the activation of a hazard and culminates in a flow of sequential and concurrent events until the system is out of control and a loss is produced. Although we think in terms of the events proceeding logically, we should remember that environmental influences are part of these logical relationships. Accident events may be fire, explosion, high energy release, destruction of parts, separation of parts of the system, and so on. We must focus on the set of events that occurs and leads to the accident, resulting in a loss. Thus, the *accident* can be defined as a dynamic mechanism that begins with the activation of a hazard and flows through the system as a series of events, in a logical sequence, to produce a loss. A simpler statement from another point of view is that an *accident* is an undesired and unplanned event that results in death, injury, or property damage. An *incident* is closely related with a definition of an unforeseen event or occurrence that does not result in death, injury, or property damage. This is commonly called a "near miss." The word *mishap* is commonly used in place of *accident*. Damage is partial or total loss of system hardware, the environment, or closely related property due to the occurrence of the accident.

1.1.5. Safety

If we return to the dictionary, we find safety defined as "the condition of being free from undergoing or causing hurt, injury, or loss." Put another way, total safety is freedom from potential harm. The system safety professional may use such terms as *increased safety*, *improved safety*, and *safer*. The difficulty with these terms is measuring *increased* or *improved* in a system that is essentially safe. Safety should be thought of as a characteristic of a system, like quality, dependability, or reliability. These characteristics are an integral part of a system (and a certain confidence in the system functions may be attributed to them by their designers and operators). *Safety* in a *system* may be defined as a quality of a system that allows the system to function under predetermined conditions with an acceptable minimum of accidental loss.

1.1.6. Risk

Risk is associated with likelihood or possibility of harm. Put another way, it is the expected value of loss. Just as the activation of a hazard can result in an accident, so the risk is related to the probability that frequency, intensity, and duration of the stimulus will be sufficient to transfer the hazard from a potential state to a loss. Risk is an expression of the possibility of a mishap in terms of hazard severity and hazard probability. Risk measurements and their relation to basic terms of hazards will be addressed later (see Chapter 32).

1.1.7   Other Hazard Definitions

The following additional definitions are used in defining hazards.

Hazardous Event: an occurrence that may lead to an accident.

Hazardous Event Probability: the likelihood, expressed in quantitative or qualitative terms, that a hazardous event will occur.

Hazard Probability: the aggregate probability of occurrence of the individual hazardous events that create a specific hazard.

Hazard Severity: an assessment of harm that could be caused by a specific hazard.

Failure is commonly associated with reliability and is related to the performance of a product or system. A system that does not perform is considered to have failed. Thus, a simplified definition of failure is:

An event, or inoperable state, in which any system or part of a system does not perform as previously specified.

For failure to be specified, the system must be defined such that the performance has been documented and the manner of operation has been carefully defined. Failure effects relate to the consequence of a failure mode on the operation, function or status of an item. Failures are classified as local effect, next higher level, and end effect. Failure will have a safety use since a great amount of the hazard data will be available from failure mode information of the product or system.

1.2.   SYSTEM SAFETY CONCEPT

1.2.1   What Is System Safety?

The system safety concept is the application of special technical and managerial skills to the systematic, forward-looking identification and control of hazards throughout the life cycle of a project, program, or activity. The concept calls for safety analyses and hazard control actions, beginning with the conceptual phase of a system and continuing through the design, production, testing, use, and disposal phases, until the activity is retired.

In the past, safety programs were usually established piecemeal, based on an *after-the-fact* philosophy of accident prevention. For example, an aviation approach is often called the "fly–fix–fly" approach: build it and fly it; if it doesn't work, fix it and try flying again (Figure 1.1). When an accident occurs, an investigation is

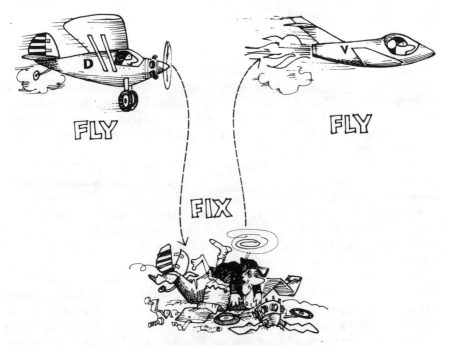

Figure 1.1 The fly-fix-fly approach to safety.

conducted to determine the cause. Accident causes are then reviewed and discussed to determine what must be done to prevent similar accidents. The resulting system modifications, retrofits, or correction of design safeguards or procedures are made to existing systems. However, corrections can be wasteful and costly and are usually vigorously resisted because of previously committed investments.

The system safety concept, on the other hand, involves a planned, disciplined, systematically organized, and *before-the-fact* process characterized as the *identify-analyze-control* method of safety. The emphasis is placed upon an acceptable safety level designed into the system prior to actual production or operation of the system. The system safety discipline requires timely identification and evaluation of system hazards—before losses occur. These hazards must be eliminated or controlled to an acceptable level to provide a system that can be developed, tested, operated, and maintained safely. Required corrective actions are identified beforehand. Proper application of the system safety concept requires a disciplined use of technical methods, including management controls necessary to assure its timely and economical completion. Engineering system safety is a vital part of the design process that must be performed on a product or system. Safety is a requirement for a Professional Engineer (P.E.) licensed by a state. This acknowledges that safety is an indispensable part of the work that is performed to qualify the design. This starts with an engineering concept. Requirements for using qualified safety professional engineers are now listed on many projects, notably the space programs for intercontinental ballistic missiles (ICBM), satellite programs and nuclear reactor building development.

The trend in product/system design is to have qualified safety personnel participate in creating that design. With the complexity of systems growing, the need for personnel giving their efforts to the safety integrity of products has increased. Most major corporations show this in their safety departments across the departmental structure with responsibilities to ensure safety of the product. Chapter 4 illustrates these organizations and their methods of building safety into major program structures to ensure the program's productive use.

Hazard analysis is the heart of the system safety approach. An effective hazard analysis effort over the life cycle of a system is the spine on which all body components of a safety program are attached. Anticipating and controlling hazards at the design stages of an activity is the cornerstone of a system safety effort. Outputs produced by properly performed hazard analyses can form the basis for specifying maintenance tasks, training, operation procedures, and estimates of the system's safety level, and also for meaningful safety reviews.

System safety is *not* failure analysis. *Hazard* is a much broader term than failure; it involves the risk of loss or harm. It is an existing or potential set of circumstances and actions that can transform an activity involving a hazardous condition into an accident or even a catastrophe. A *failure*, on the other hand, is when something functions in a manner in which it was not intended. A failure can occur without loss. Severe accidents have happened while something was operating exactly as intended—that is, *without* failure.

Engineering approaches are also different. For example, increasing the safety factor of a hazardous material containment system by increasing the burst pressure/ working pressure ratio of a tank introduces new dangers in the event of a rupture resulting in fire explosions or chemical reactions. Hazard analyses looks at these interactions, not just at failures or engineering strengths. Hazard analyses may include "failure mode and effects" analyses, but they go beyond this one analytical method. System safety analytical methods are specialized *safety* analysis methods, involving *safety* disciplines, principles, and procedures. They are normally practiced by professional *safety* analysts.

1.3. SYSTEM SAFETY HISTORY

In September 1947, a technical paper titled "Engineering for Safety" was presented to the Institute of Aeronautical Sciences. It stated the following:

> Safety must be designed and built into airplanes just as are performance, stability, and structural integrity. A safety group must be just as important a part of a manufacturer's organization as a stress, aerodynamics, or a weights group.

This paper provides one of the earliest recordings of the system safety concept. It was not until the early 1960s that the concept was applied formally by contractual direction. This formal delegation of safety responsibility by contractual requirement replaced the familiar practice in which each designer, manager, and engineer presumably assumed his share of the responsibility for safety. The growth and development of the system safety approach to accident prevention is created by the publication of safety standards, specifications, and requirements, as well as operating instructions.

The missile systems developed in the late 1950s and early 1960s demanded a new approach to examining the hazards associated with the weapons systems. The Minuteman intercontinental ballistic missile (ICBM) was one of the first systems to have a formal, disciplined, system safety program associated with it. Previously the Atlas and Titan ICBM had safety design examinations, associated chiefly with the main engineering design conducted by the prime contractors. Colonel George Ruff, assigned to the Ballistics System Division of the U.S. Air Force, was one of the leaders who participated in initiating the first system safety program required of the prime contractors on the missile programs.

In April 1962, the Air Force published BSD Exhibit 62-41, "System Safety Engineering for the Development of Air Force Ballistic Missiles," applicable to Ballistic System Division programs. This document established system safety requirements for the associate contractors on the Minuteman missile program, where the first real system safety groundwork was done. In September 1963, the document was revised into Air Force specification MIL–S–38130, "Military Specification—General Requirement for Safety Engineering of Systems and Associated Subsystems and Equipment." With very minor revision, in June 1966, this specification was made a Department of Defense requirement, MIL–S–381308A. In 1969, the specification was revised further and became MIL–STD–882, "System Safety Program for Systems and Associated Subsystems and Equipment; Requirements for." Department of Defense approval of MIL–STD–882 then became mandatory for a system safety program on all procured products and systems. In 1984, MIL–STD–882B was promulgated. A more specific tailored program was now in place.

Just as the Air Force gradually developed system safety requirements for contractors to meet, so the National Aeronautics and Space Administration (NASA) recognized the need to have system safety as part of their programs. Successful Air Force programs provided valuable data on the hazards of parts and systems and the methods of hazard control. As a result, several initiating documents were incorporated by NASA through Jerome Lederer, head of the NASA Safety Office. Through his leadership, a large task force of systems engineering staff was developed, as was an extensive program of system safety for space activities. Much of the success of the NASA programs can be attributed to the effort that system safety played in the hazard identification, evaluation, and control.

Early recognition of safety analysis needs was brought to the public by Chauncey Starr, Vice-President, Atomics International Division, North American Aviation, in the 1960s. During successful space ventures, the technology transfer was considered a very useful aspect of the space program. The contributions in safety were universal as a result of the many years of space application. NASA conducted regular government industry system safety conferences in 1968 and 1971 to assist in the dissemination of information concerning the latest system techniques and methods to safety professional personnel.

In a similar manner, the use of system safety programs have grown in commercial industry, as programs pioneered by the military were adopted. An important, and perhaps the most sophisticated, use of system safety methods today is in the nuclear power, refining, and chemical industries. The current state of system safety has been through a progressive increase in definition and focus by use of the MIL–STD. With the growth of systems in complexity, the safety standard has increased its ability to provide the tools that can be used in applying system safety to the product or system. This growth is seen in the need for tailoring the safety standard for application and the need to specifically tie down the requirements to deliverables that are important for product development.

An example of this growth is seen in the addition of software system safety to the MIL–STD–882. Seven tasks of definition are now needed to understand clearly the requirements for considering how software is used with the total program. The interfacing of hardware, software, and firmware has made it a necessity to define these additional tasks and apply them carefully in the system safety developed program. This is seen in the large space programs where software is used to control space vehicles. The operator observes and makes critical decisions, but, for the most part, relies on the software for vehicle control. As a result, hazards identified in the software must be examined before the safety integrity of the vehicle (with its hardware, software, and firmware) is released for production.

The outline for a system safety program, as detailed in MIL–STD–882 is used by many industries in establishing a corporate system safety program for the products they are developing. Requirements for standards evolving from the American National Standards Institute (ANSI), American Society of Mechanical Engineers (ASME), and many other professional societies have culminated in a concerted effort to insure that system safety principles are applied to products used by the public. System safety has come of age in many bases of commercial design and manufacture. The lessons learned are that system safety pays off in effective products, fewer accidents, and a longer product life.

1.4. SYSTEM SAFETY OBJECTIVE

A safety objective such that each person will live and work under conditions in which hazards are known and controlled to an acceptable level of potential harm.

The tasks needed to achieve the objective above require recognition that the rate at which technology is changing is accelerating constantly. Since this rate of change is a function of access to available knowledge, recent innovations in computer technology alone assure exponential growth for the foreseeable future. In turn, this acceleration of technological growth:

1. *increases the number of individuals who may be exposed to a given hazardous incident.* Consider, for example, the rate at which passenger capacity has grown in aircraft.

2. *establishes a need to cope constantly with safety problems for which no previous experience can serve as a guide.* Consider, for example, the recent accumulation of knowledge concerning the harmful effects of toxic substances.

1.4.1. Procedure for Achieving the Objective

The tasks that must be carried out to achieve the objective fall, with some degree of overlap, into two general categories. First, the traditional set of tasks are oriented toward establishing standards, making products safer, and establishing a safer climate for product manufacture and use. This traditional approach is discussed by Haddon, Peterson, and Grimaldi, among others. The second, system safety, only a few decades old, is more quantitative in nature and is rooted in systems and operations research technology. Its early military applications included, for example, assuring that inadvertent nuclear explosions would not occur and that space travelers would be safe in their journeys. System safety is now used in a multiplicity of domains— mass transportation, petroleum production and distribution, nuclear power plant construction, and chemical facility design.

In essence, this second category is concerned with determining an optimal degree of safety, within the constraints of operational effectiveness, time, cost, and other applicable interfaces to safety, that can be achieved throughout the life cycle of the system. Like the traditional approach, system safety is concerned with controlling safety while taking other factors into consideration. For example, it is frequently necessary to quantify variations in the safety level of a system as a function of defects in design, material and workmanship, human errors of omission and commission, and interfaces of the system with the environment and other equipment over the life of the system.

The goal of a system safety program is the elimination or control of hazards. This will reduce the potential loss of a system, reduce the potential injury or morbidity, and reduce the potential damage to the system or related equipment to an acceptable level. This goal can be achieved at minimum cost when the system safety program is implemented early in the concept phase and is continued throughout system development and operation life cycles.

The complexity and involved interrelationship of elements within, and external to, a system require detailed system safety studies. Potential hazards are detected and the probability of occurrence is estimated. The phases include normal operational modes, maintenance modes, failure modes of the system, failures of the adjacent equipment, and errors created by human performance. Another often-stated goal of a system safety program is the *creation of a reasonably safe product* (Weinstein et al. 1978).

The system safety process, simply stated, analyzes each component and operational procedure of a proposed system, for its hazard contribution. It first identifies, then analyzes potential hazards so that action can be taken to minimize their unsafe effects. It can also alleviate *design-induced maintenance* and *personnel errors*. The next step in the process is to eliminate the hazard by design or minimize its effects by warning devices, revised operating procedures, or other effective means. The ultimate objective, again, is to identify and then eliminate or control all hazards.

In the final analysis, safety engineering and management judgment are the key factors in the determination of acceptable risk. The factors considered are evaluation of cost, probable effectiveness, probability of damage, notoriety factor, frequency of exposure, and balance of system benefits against possible losses. Mathematical approaches can aid but should not be considered the only means of solution.

The attempt is to *design out* most of the potential accident causes while the system is still on the drawing board. The system safety engineering concept was formed on the premise that there was a real need for more safety awareness and emphasis early in the design process. This premise is still valid.

All known facets of safety program characteristics—including design, engineering, education, management policy, and supervisory control—should be considered in identifying and eliminating or controlling hazards. System safety management and engineering should be integrated with other management and engineering disciplines in the interest of optimal system design. Procedures for development and integration of the system safety effort should be applied across the managing activity considering the three principal elements of the system process that must be blended into a safety system (see Figure 1.2) design.

The managing activity should establish, plan, organize, and implement an effective system safety program that is integrated into all life cycle phases. The responsibility and functions of those directly associated with system safety policies and implementation of the program should be clearly defined, assuring that all safety inputs

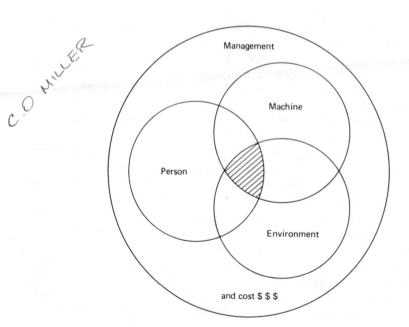

Figure 1.2 Elements of system design process.

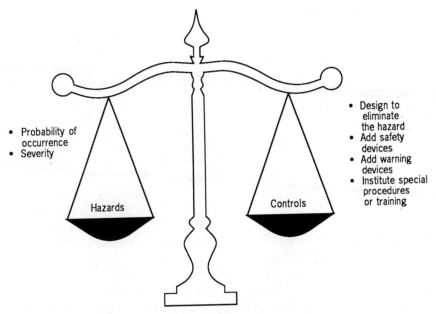

Figure 1.3 Tradeoffs of hazard management.

to program milestones and reviews will be made. To ensure compliance, a system safety program plan (SSPP) should be a contractual requirement.

The system safety objective can be thought of as the balance between risk and controls. The management of this balance is illustrated in Figure 1.3, where the factors of risk (probability of occurrence and severity) are weighed against the controls required to balance these risks. The controls depicted are those fundamental methods by which hazards are controlled prior to the occurrence of the accident. These controls are discussed more thoroughly in Chapters 2 through 6.

Hazard severity is a key factor in understanding the first part of the safety program goal. MIL–STD–882 has defined four suggested categories of hazard severity: Category I, catastrophic; Category II, critical; Category III, marginal; and Category IV, negligible. Table 1.2 depicts these categories and provides a means to place the hazard results into one of these categories. Further definition must be made in the delineation of the words "severe," "major," and "minor," with some degree of quantitative values assigned to the type of category. This definition would be performed with the assistance of past data on similar programs and a careful analysis of the various hazard conditions that could result.

In the case of aircraft, the differences between major and minor damage can be defined in the amount of dollar value of loss, the type of loss affecting the system (such as engine, cabling, radar, etc.), or the maintenance action required, such as number of days required for repair (number of days of down time). In the case of personnel, the difference between "severe" and "minor" injury should be carefully defined. (This difference is a function of the worst case potential injury resulting

Table 1.2. Hazard severity

Category	Name	Characteristics
I	Catastrophic	Death
II	Critical	Loss of system Severe injury or morbidity
III	Marginal	Major damage to system Minor injury or morbidity
IV	Negligible	Minor damage to system No injury or morbidity No damage to system

from the hazard.) Thus, it is obvious that these definitions are tailored to become more specific with the particular program and are made with some degree of understanding of the types of hazards.

Finally, with respect to the probability of occurrence, a table has been established in the MIL–STD–882 that defines a scale of likelihood (Table 1.3). The categories of likelihood (classes A to E) allow the specific event to be rated as to its likelihood of occurrence. The method of determination may be a reliability value (failure rate) in the case of a mechanical or electrical device. (Many times these failure rates are actually available from tables in MIL–HDBK–217E, trade association manuals, or published manufacturing data.) However, there will be difficulty in assigning values, particularly when the failure definitions for new items are not known.

The qualitative hazard likelihood ranking, which provides a rating scale for hazards and is used later in risk definition to determine the items that must be controlled, is important. Redundancy in design has been a major factor in providing a less probable value for the hazardous item. However, design redundancy should

Table 1.3. Possible hazard likelihood

Description[a]	Level	Specific Individual Item	Fleet or Inventory[b]
Frequent	A	Likely to occur frequently	Continuously experienced
Probable	B	Will occur several times in life of an item	Will occur frequently
Occasional	C	Likely to occur sometime in life of an item	Will occur several times
Remote	D	Unlikely but possible to occur in life of an item	Unlikely but can reasonably be expected to occur
Improbable	E	So unlikely, it can be assumed occurrence may not be experienced	Unlikely to occur, but possible

[a] Definitions of descriptive words may have to be modified based on quantity involved.
[b] The size of the fleet or inventory should be defined.

Table 1.4. Hazard Assessment Matrix

Frequency of Occurrence	HAZARD CATEGORIES			
	I Catastrophic	II Critical	III Marginal	IV Negligible
(A) Frequent	1A	2A	3A	4A
(B) Probable	1B	2B	3B	4B
(C) Occasional	1C	2C	3C	4C
(D) Remote	1D	2D	3D	4D
(E) Improbable	1E	2E	3E	4E

Hazard Risk Index	HRI	Suggested Criteria
1A, 1B, 1C, 2A, 2B, 3A	I	Unacceptable
1D, 2C, 2D, 3B, 3C	II	Undesirable (management decision required)
1E, 2E, 3E, 3E, 4A, 4B	III	Acceptable with review by management
4C, 4D, 4E	IV	Acceptable without review

be analyzed carefully to understand the actual impact on safety since improving reliability can sometimes lessen the safety level of the system or product. Table 1.4 illustrates the method to establish a hazard risk index (HRI) with the use of frequency of occurrence and hazard category. Table 1.5 illustrates the relationship of event occurrence level to a quantitative value. First as a low, medium, or high condition which is frequently used with actual values are obtained.

Table 1.5. Relationship of qualitative probability rankings to quantitative values

Description	Level	Frequency of Occurrence	Potential Relationship to Quantitative Value[a]
Frequent	A	High	10^{-1}
Probable	B	↓	↕
Occasional	C	Medium	$>10^{-3}$
Remote	D	↓	$>10^{-4}$
Improbable	E	Low	$>10^{-6}$

[a] All quantitative values require a database for establishing the value.

1.5. SYSTEM SAFETY AS A DESIGN PARAMETER

Operational suitability and safety of a system in its service use is dependent on several highly integrated task requirements (see Table 1.6). The service–use factors, in combination with the functioning of the system, provide a simple measure of how useful the system is in its operational role. Therefore, it is necessary to develop a total program in which design analyses, studies, and testing will identify limits on performance requirements, hazardous failure modes, reasonable safety margins, and recognition of critical operations. Identification of these potential problem areas permits system safety design and operational criteria to be established for hazard elimination or control. This approach forces acknowledgement that a specific safety problem cannot be treated without regard for its interfacing consequences with related elements within or without the system. Interactions between all design support disciplines (such as safety, human engineering, reliability, and maintenance) must be controlled and properly channeled to achieve a system that is responsive to all requirements.

System safety engineering employs techniques of system engineering to analyze a system as an interrelated set of interacting elements generating hazards. Detection, elimination, and control of hazards is accomplished through use of proved techniques and well-established safety practices.

The following examples illustrate the penalities of poor system safety design practices.

1. In a recently designed aircraft, a fuel (JP-4) vent was located in proximity to the wheel brake area without redundant float valves having been designed in the vent line. A single failure of the vent float valve would allow fuel to ventilate onto hot wheel brakes, a situation which could easily lead to fire. Redundant or backup systems would help reduce the probability of this type of mishap.

2. On a two-engine aircraft, the landing gear was designed to be raised by using hydraulic pressure from a pump on the number one engine. Should that engine fail on takeoff, the pilot would need all the air speed he could muster from the remaining engine. With the landing gear still down, the drag would be

Table 1.6. System safety task requirements

Safety management
Safety engineering
Safety analyses
Hazard identification
Hazard description
Cause determination
Hazard control
Control evaluation
Documentation

considerable and could prove disastrous. New designs permit the gear to be raised by hydraulics from more than one engine.

3. A solid-state electronic system was designed to protect nuclear reactors. A nonsolid-state system of similar design had performed flawlessly for years. The solid-state system was tested for a year and then installed on a high-flux isotope reactor, where it seemd to perform without failure and facilitated daily testing of the system. Some years later the same system was installed on a fast-burst reactor. On the first burst it was found that the overload destroyed the solid-state components essential for the system's functioning. Previous reactors with the solid-state system had, in fact, not been protected and the system had responded only satisfactorily to test loads.

Examples of this sort are not limited solely to the systems described but they illustrate design deficiencies that system safety techniques can detect. The costs of retrofit or modifications are extremely high once the system is in production. The cost of litigation, if a serious mishap were to occur, can also be extremely high.

Over the last two decades the Nuclear Regulatory Commission has established very strict controls on exposure to nuclear material. Their safety controls have been directed toward elimination of undesirable events with emphasis to the design features of the system to maintain safe control.

C. O. Miller (1980) reviewed the historical development of safety in the nuclear power system area. He noted that the manner in which the safety design and safety validation checks were performed was basically a random engineering participation by many divisions of the Nuclear Regulatory Commission. Since the Miller report, major improvements in the NRC safety program, meant to bring forth a more integrated plan that covers aspects of system safety fully, are being accomplished. This is a continuous process.

1.5.1   MIL–STD–882

This standard provides uniform requirements for developing and implementing a system safety program that will identify the hazards of a system and ensure that adequate measures are taken to eliminate or control them. This standard applies to the Department of Defense (DOD) systems and facilities, including support, test, maintenance, and training equipment. It applies to all phases of the system life cycle. The principal objective of a system safety program within the DOD is to ensure that safety consistent with mission requirements is designed into systems, subsystems, equipment, and facilities.

Part of the MIL–STD–882 reads as follows:

> This standard provides uniform requirements for developing and implementing a system safety program of sufficient comprehensiveness to identify the hazards of a system and to impose design requirements and management controls to prevent mishaps, eliminate hazards, or reduce the associated risk to a level acceptable to the managing activity.

MIL–STD–882 is built around the development of a program tailored to the detection of hazards for the type of system that is under design consideration. Examples in the task listing for the system safety program illustrate that the first effort is to decide the *complexity* of the product or system to be designed. When the complexity is defined, then the tailoring of the selection of the tasks for the program can be made for appropriately adding only the required analyses efforts for use in the program. For example, the use of Sneak Circuit Analysis (SCA) is an analytical tool that can be selected. It usually requires a computer to compare electrical wiring within a product/system to understand the unusual forms of hazards that can result from the performance of the system. Because of an increase in electronics in control of equipments, this type of analysis can be very costly and requires many hours to complete effectively. Therefore, the use of this task will require an acknowledgment of the need for conducting such a lengthy analysis.

Some of the considerations in the selection of system safety tasks will be evident after a clear definition of the type of system that is to be developed. For example:

1. Software safety tasks are not needed unless there is software development in the system.
2. Certain tasks will not be done unless they correspond to the time in the life cycle when the system safety work would be required.
3. The use of specific analytical efforts such as Fault Tree Analysis (FTA) will require a definition of the depth to which the FTA should be taken. If this is not done, difficulty in achieving a useful result of the fault tree can result.
4. Safety verification testing will require definition after the identification of hazard controls and decision of the specific types of testing to be conducted.

MIL–STD–882 has been deliberately revised to assist the user to make tailoring easy and selecting tasks routine. Only definition of the system and detail of work are the key items demanding attention.

2

SYSTEM LIFE CYCLE

System safety is a discipline designed to affect the total life cycle of a product or system. We have described several system safety programs that have been active in the last several decades. The common theme among these programs has been the continued active performance of system safety tasks over the life of the system. The focus on hazards from the system's beginning through the evaluation and control of the hazards is the core of the program. To understand the particular tasks to be performed and to assure that they are accomplished in an orderly and logical manner, a division of the cycle is defined.

2.1. LIFE CYCLE DEFINITIONS

The five common phases of a life cycle are concept, definition, development, production, and deployment. These phases correlate to nomenclature established for system evolution that can be seen in governmental and industrial practices. The phases selected are shown related to other nomenclature in the following chart.

Selected Phases	Other Nomenclature
Concept	Conceptual research
Definition	Validation
Development	Full-scale engineering development
Production	Manufacturing
Deployment	Operation—maintenance

Life cycle phases are important divisions in the evolution of a product. The ability to perform specific tasks in each portion of the evolution is the key to the successful control of hazards.

Disposition (also known as termination) is a sixth phase of the life cycle. This is the retirement of the product or system. There is good reason to acknowledge this phase and understand that system safety design effort must also be directed to the safe disposition of hazardous or toxic products (Section 6.7 discusses this in more detail).

An overall survey of hazardous effects associated with the product should offer information on all hazards in each phase of the life of the product. Brown's safety examples (Brown 1976) emphasize the importance of design for safety as a part of the overall effort to control loss in each portion of the life cycle.

2.2. SYSTEM SAFETY CONTROL MILESTONES

There are five major control points for system safety in the development of a system; they are the end points of the phases of concept, definition, development, production, and deployment. Table 2.1, the design review of system safety activities, shows end points at which the design presented must either be accepted or stopped for examination prior to proceeding to the next phase. The specific design review points are the following

Phase	Safety Control Point	Result
Concept	Concept design review	Establish basic design for general evolution
Definition	Preliminary design review	Establish general design for specific development
Development	Critical design review	Approve specific design for production
Production	Final acceptance review	Approve product for release in deployment
Deployment	Audit of operation and maintenance	Control of safety operation and maintenance

It is important to recognize these design review areas. It is just as important to insure that sufficient time is spent on the review process. A clear understanding of all data presented at the design review time is necessary before the system is allowed to enter the next phase. Table 2.2 lists the system safety data item description (DID) items for the contract requirements. These items are standard and assure deliverable data for the items in Table 2.1. The value of the review process cannot be emphasized enough (Childs 1979), as is shown by the number of defects and oversights that commonly enter the operational phase. The costs incurred at that time can be *two* to *ten* times the costs resulting from changes in a proper review.

2.3. CONCEPT PHASE

The concept phase is the initial period in which historical background data and future technical forecasts are developed to provide a basis for the proposed system hazard analysis. Critical issues related to the product are examined, system safety concerns with types of hazard are identified, and their impacts are evaluated. A preliminary hazard analysis (PHA) is an analytical tool used during the concept phase to bring out the hazards that would be involved with a specific concept. Risk analysis (RA) on a gross level would also be performed to determine the immediate needs for hazard control and the development of safety design criteria.

System safety management will be oriented to the development of a system safety program plan (SSPP), identifying the tasks to be accomplished in the total safety program for the evolution of the system. The effort to develop a SSPP at this time is of major importance to assure that safety is examined in a logical, sequential manner throughout the entire program.

The PHA is performed as the initial hazard analysis in the development of the system. Risk analysis involving techniques described in Chapter 28 are also used in a preliminary form to reach conclusions about the safety of the design.

Biancardi and Peters have both stressed the importance of the concept phase of a safety program in enlightening the producer as to the potential hazards that are a part of the final design (Biancardi 1971, Peters 1965). The recognition of these potential hazards is the first task. Nonetheless, the challenge of identifying the control method that will result in an acceptable *deployed* system remains. A special emphasis is placed on hazards that can be generated by energy sources.

Three basic questions should be answered at the close of the concept phase.

1. Have the hazards associated with the design concept been discovered and evaluated to establish hazard controls?
2. Have risk analyses been initiated to establish the means of hazard control?
3. Are initial safety design requirements established for the concept so that the next phase of system definition can be initiated?

2.4. DEFINITION PHASE

The definition phase provides for verification of the preliminary design and engineering of the product. The SSPP should be specific in the safety tasks that are to be undertaken during this phase, particularly the identification of analyses that should be conducted. Models may be developed to allow more comprehensive investigation into the hazards of the product and identification of the preliminary design. Areas of technological risk, costs, human engineering, operational and maintenance suitability, and safety are examined in detail and reported at design reviews in this phase.

Table 2.1. System safety activities

	PHASE OF TOTAL PROGRAM						
	R & D	ACQUISITION			OPERATION	TERMINATION
Control Points	CDR₁	PDR	CDR₂	FAR		
Safety Activity	Concept	Definition	Development	Production	Deployment	Disposition
1. Formulate system safety program plan	▢					
2. Develop safety design criteria	▢					
3. Conduct hazard analysis	PHA	PHA/SSHA/SHA/SWHA				
4. Define safety design requirements	Initial	Final				
5. Conduct design reviews	In SSPP Schedule					
6. Provide safety inputs to manuals		Submit and review			Review	

24

7. Participate in failure analysis

8. Conduct risk analyses

9. Conduct documentation reviews

10. Identify and define safety equipment

11. Prepare safety test plans and requirement

12. Conduct safety tests

13. Conduct safety training

14. Participate in accident investigations

Task				
7. Participate in failure analysis	Historical data review	Test results		
8. Conduct risk analyses	As required	As required		
9. Conduct documentation reviews			As required	
10. Identify and define safety equipment	Initial	Preliminary	Design	
11. Prepare safety test plans and requirement	Initial	Preliminary		Update and verify
12. Conduct safety tests		Prototype	Proof	Acceptance
13. Conduct safety training			Support	Monitor
14. Participate in accident investigations		As required		Reverify

[a]PHA: Preliminary Hazard Analysis; FTA: Fault Tree Analysis; CDR$_1$: Concept Design Review; PDR: Preliminary Design Review; CDR$_2$: Critical Design Review; FAR: Final Acceptance Review; SSHA: Subsystem Hazard Analysis; SHA: System Hazard Analysis; SWHA: Software Hazard Analysis.

Table 2.2. System safety DIDs for contract requirements

Title	Items Covered
System Safety Program Plan	Task and activities of system safety management and engineering to identify, evaluate, and eliminate or control hazards in life cycle
System Safety Hazard Analysis Report	Hazard Analysis reports Preliminary Hazard Analysis (PHA) Subsystem Hazard Analysis (SSHA) System Hazard Analysis (SHA) Operating and Support Hazard Analysis (O&SHA)
Safety Assessment Report	Comprehensive evaluation of safety risks being assumed prior to test or operation of the system
Engineering Change Proposal System Safety Report	Summarize results of analyses, tests, and trade-off studies on proposed design changes in system life cycle
Waiver or Deviation System Safety Report	Summarize results of analyses, tests, and trade-off studies as they relate for waiver/deviation
System Safety Program Progress Report	Periodic review of safety activities and monitoring of progress of contractor's system safety efforts
Occupational Health Hazard Assessment Report	Identification and evaluation of health hazards and proposed measures to eliminate or control hazards through engineering design changes or protective measures

To provide a design that meets the criteria derived from the concept phase, a clear definition of the subsystems, assemblies, and subassemblies of the system must be made. An examination of the hazards of several designs may be required. An updating of the PHA is accomplished, along with initiation of the subsystem hazard analysis (SSHA) and later integration into the system hazard analysis (SHA). The fault hazard analysis (FHA) and fault tree analysis (FTA) may be used to examine specific known hazards and their potential effects. (Refer to Table 2.1 for performance periods of major system safety activities.) Risk analysis will help evaluate the different hazards that may be considered in the preliminary stages. An examination of the risk is the key to selection of the final design.

One or more safety analysis techniques may be needed to identify the following: safety equipment, specification of safety design requirements, initial development of safety test plans and requirements, and prototype testing to verify the type of design selected. Understanding the need for a complete evaluation of hazards to assure that controls are considered in the preliminary design of the product or system is vital. System definition will initially result in acceptance of a suitable general design, which must then be more specifically developed in the following phase. Not all hazards will be known at this time, since the design is not complete.

2.5. DEVELOPMENT PHASE

The development phase allows system definition to include environmental impact, integrated logistics support, producible engineering, and operational use studies. Prototype analysis and testing results are used as inputs for a comprehensive O&SHA to examine human-machine hazards. Further, an SSHA will be developed, since the specific design will now be presented in a more complete form. The safety design criteria should be complete and provide the detailed information for hazard control of the system.

More complete testing is performed in the development phase to allow a verification of the acceptability of the design. Any failures should have been examined and analyzed for their safety impact. Appropriate action may be taken by correction of the design, warnings, procedures, or training. Also, the results of the fault tree analysis should have verified the acceptability of major undesired events.

Interfaces with other engineering disciplines will have been exercised in this phase, particularly reliability engineering with the review of the failure modes and effects analysis (FMEA). Failure modes that are hazardous should have been clearly identified in the hazard analysis, and action should have been taken for their control. The FMEA will be contributing information for the SSHA and FTA.

The completion of the development phase should lead to a go/no-go decision on a specific design before production begins. The ability to make the correct go/no-go decision is based upon completion of hazard analysis, safety testing results, and complying with safety design criteria. If any aspect of the safety review were to be neglected, costs will probably be inflated. A poor safety review will always lead to a higher product cost.

2.6. PRODUCTION PHASE

Monitoring by the safety department during the production phase is most important. With production commencing, the quality control (QC) department serves as a major focal point for inspection and testing of the product. Therefore, interacting carefully with QC is absolutely necessary. Depending on the complexity of the product, it may be desirable to have a qualified system safety person assigned to Manufacturing to witness tests that involve safety and to verify the conduct of critical tests leading to the final acceptance of the product. In areas where qualification tests are conducted, this witnessing is necessary since the check upon safety devices and their proper functioning should be done only with a responsible system safety person present.

Training will be initiated during this phase. It is quite important for the safety person to monitor the total training program to assure that safety training is occurring. The training is established so that operators and maintainers may work with the actual equipment. Later, in delivery to the customer, this same training is repeated to assure that the training incorporates the results of the O&SHA for proper hazard control in operation and maintenance.

The updating of the analyses performed during the definition and development phases will occur during this phase. An objective review of past hazard analyses to verify that corrective action for hazards has been incorporated in the manner set forth in the documentation is a requirement. Any changes that occur will be subject to review and verification during the final acceptance review (FAR).

Finally, the system safety engineering report (SSER) is a compilation of the production phase inputs that identifies and documents the hazards of the final product or system. This report should describe the safe use of the product in the environment in which it may be deployed. It represents the data obtained from the analyses, testing, and design criteria evolution. The SSER should provide definite conclusions about the safety integrity of the product and the manner in which specific hazards identified have been controlled. It should also show the manner in which the operational deployment of the product can be accomplished for the customer. For example, NASA has demonstrated in its programs the positive use of system safety engineering reports and the value of careful design reviews prior to the use of their products. NASA's early checks of the design of the production system, combined with careful safety testing have resulted in useful, safe products. The emphasis on the SSER was a major factor determining the success of these programs.

2.7. DEPLOYMENT PHASE

The deployment phase follows production. At this time the system becomes operational. Training is conducted and data are accumulated (from production failures, field failures, and accidents and incidents that have occurred). System safety management has to be available to follow up on any problem that may arise during this period, and a system safety person should participate in the work of the accident investigation board so that identification of hazardous conditions will be made as soon as possible and corrective actions can be worked out with designers and reviewed by responsible safety personnel.

As the design group takes action for engineering changes on the product after deployment, it is necessary that the system safety person have the opportunity to review design changes that may be submitted. It is necessary to assure that hazards are controlled and that no new problem is introduced into the system as a result of an engineering change. At times an additional preliminary hazard analysis (PHA) or an updating of some specific hazard analysis that has been performed previously on the system may be required.

Finally, a sixth phase—termination—may be significant because of certain hazardous elements of the system. The system safety person should be available to check out the previously developed procedures for the product termination and to verify that the method is carefully monitored. Since the actual termination occurs at the end of operation, this monitoring should be performed on a sampling basis to verify the correct use of the termination procedure where it involves a hazardous situation or substance.

3

SYSTEM SAFETY IMPLEMENTATION

Implementation of a system safety program centers on the commitment by a company's top management. When this commitment is made a system safety policy can be developed through the president's or chief executive officer's action. In most industrial organizations the responsibility for conducting a system safety program is assigned to the project management level. Project management is charged directly with the responsibility for the development of a product or system. Management is then required to plan and implement a proper program to accomplish the goal of a safely designed product. Military safety requirements have been set forth by each service as well as in the U.S. Department of Defense MIL–STD–882 document.

3.1. POLICY AND PROCEDURES

Where contracturally required, the system safety program involves establishing safety policies for each project. Also, for projects where no specific contract requirement for system safety exists, company management may wish to designate a planned system safety program (for reasons such as unusual company liability exposure or unusually hazardous nature of the product). Some of the major program objectives in such a plan are the following:

1. *Safety is consistent with the requirements of the system.* Examination of previous programs will provide the means to ascertain that proper consistency of design is maintained. In the case of repeating designs, the clear call-out of safety design requirements in a manner proved in previous applications is appropriate. If newer information is available, the design approach will be altered to provide changes that would be useful to control hazards. An example is the automobile. For many years the standards of safety application were the same, with, only the

body changing yearly. Presently, we see many new automobile safety standards applied to bring about increased safety of design with advanced technology such as microprocessors, antiskid braking systems, and improved restraint systems.

2. *Reasonable risk is involved in the acceptance and use of new materials and new production and testing techniques for the product.* The methods used to analyze new materials in production and testing involve a detailed analysis of the trade-off, between risk and use of these items. It is necessary to evaluate risk of the hazards and at the same time consider their controls to facilitate an evaluation of the acceptance level of risk.

3. *Hazards associated with the product or system are identified and eliminated and/or controlled to an acceptable level to protect personnel, equipment, and property.* Close coordination is required with the industrial safety personnel of a company to insure that protection of personnel, equipment, and property has been properly accomplished. The interaction between the people responsible for system safety and industrial safety is of great importance. The worker and environment have significant roles in determining acceptable levels of hazards. Close examination of environment standards and codes along with an examination of the worker tasks are necessary to determine acceptable levels of protection. Hammer (1976) emphasizes the close coordination required between system safety and industrial safety.

4. *Engineering changes required to improve safety are minimized through the timely inclusion of safety controls during the design and development of the product.* The method by which safety controls are brought into existence in the course of product development must be stated in a clear, positive policy. Use of material review boards, design review actions, program management reviews, and system safety working group meetings contributes to the effectiveness of this policy. These areas of action should be noted and assigned to the responsible project management personnel. Product assurance management must be aware of the necessity of using these coordination tools to initiate engineering changes. Methods of evaluating cost trade-offs caused by expected safety changes are described in "Life Cycle Cost of Safety Estimation" (Moriarty 1977).

5. *Historical safety data generated by similar system programs are considered and used where appropriate.* The use of data from other programs is most valuable. The justification and rationale for decisions will often be based upon previous actions in similar programs and the ability to apply similar techniques where the application fits. As a result, the increased value of keeping safety data and insuring its use in the future is a *hidden asset* that can assist in the decision-making process for safety requirements and eventual design action.

A minimum listing of company policies and procedures would include the following:

1. *System Safety (SS) Program.* This should establish the policy and responsibility to define and implement SS activities through every group in the company.
2. *System Safety Handbook (SSHB) Authority.* This creates the usage of a System Safety Handbook and the process for requesting and approving deviations from these requirements.

3. *System Safety (SS) Program Plans.* Develops policy, responsibility, and procedure for SS program plan activity on every program and project throughout the company.

4. *Program/Project System Safety Program Management/Engineering.* This establishes the minimum activities, responsibilities and procedures that comprise a program/project.

5. *Subcontractor System Safety.* This creates the requirements for flow down and implementation of SS requirements to subcontractors supporting the company projects, products or services.

6. *Hazard Analysis and Reporting.* This establishes the policy and procedures for performing and documenting hazard analyses.

7. *Safety Compliance.* This details the policy and procedure for verifying compliance with design and operations SS requirements, and qualitatively assessing mishap risk prior to use of a product.

8. *Operational Safety.* This establishes safety of personnel, products and government-furnished equipment through operational safety surveillance and SS engineering support during integration, test and operation, and servicing of products.

9. *Mishap Reporting.* This establishes the reporting and corrective action regarding mishaps that occur in the manufacture, assembly, and test of the products of the company.

A company should direct project management to implement a detailed system safety program. The following system safety efforts should include tasks that are integrated within the project organization to accomplish the purpose of the program objectives.

- Prepare system safety program plans. Monitor project compliance with approved plans.
- Establish safety design criteria pertinent to the product design and mission. Analyze the design through all phases to identify hazards.
- Review design disclosure documentation to assure that detailed requirements are responsive to safety criteria.
- Participate in design reviews of the projects, emphasizing safety review. Conduct risk trade-off studies in safety areas.
- Determine specific system safety objectives and goals.
- Review planned demonstration testing.
- Review test data for safety performance analysis.
- Coordinate safety support roles of design, project management, integrated logistics support, assembly, test and checkout, and other functions as required.
- Negotiate and monitor vendor safety programs.
- Coordinate project system safety plans with the industrial safety department when such plans require this support.

- Represent the project in all customer safety working group meetings. Obtain approval for close-out of safety action items.
- Investigate accidents and incidents (near misses) involving project equipment.
- Communicate with other projects regarding system safety experience and knowledge.
- Identify off-site safety regulations and laws having an impact on the project.
- Advise the industrial safety department regarding accident and incident information and reports relating to occupational safety and health.
- Document those project decisions and actions having potential influence upon related product liability exposure.

In order to implement a system safety program within a company, there are several steps necessary in order to obtain full usage of an effective program. The following process is recommended:

Process for Implementing a System Safety Program

1. Obtain top management approval for adoption of the concept.
2. Determine appropriate staff and schedule requirements.
3. Coordinate and solicit cooperation from other departments.
4. Develop and document the system safety program.
5. Obtain top management approval for the system safety program plan.
6. Distribute the system safety program plan and obtain input from other departments.
7. Implement the system safety program.
8. Assess the program and revise as necessary.

These steps are simple and straightforward. They should have the concurrence of the top level executive before obtaining the support of the other members of the organization. In order to sell a company's capability in system safety, it is important to have a system set up in your own organization that represents an effective, well-laid-out comprehensive program responsive to the needs of your company and customers. The benefit is seen as instilling confidence into the customer of your capability to lead a system safety program that will perform to the needs of associated industry requirements or military standards. A sample of chief executive officer (CEO) policy elements is shown in Figure 3.1.

3.2 PRODUCT ASSURANCE ORGANIZATIONS

During the course of development of system safety as a discipline the areas of reliability, maintainability, human factors, and quality assurance were found to be a major source of data for system safety. These disciplines hold a common link

---

CHIEF EXECUTIVE OFFICER'S POLICY STATEMENT

CEO Policy Statement on System Safety Program

The (company name) was organized to provide effective, reliable, clean, courteous, and *safety* program for our company and our customers. The safety portion of our task is of paramount importance. All (company name) personnel and appropriate contractors are charged with the responsibility of ensuring the safety of our product, employees, and others who come in contact with the product as well as the property itself.

Further, the System Safety Department is directed and empowered to devise and administer a comprehensive and coordinated System Safety Program with a specific safety plan, and the activities to prevent, control, and resolve unsafe conditions which may occur during both design and operational stages. This authority includes the right to stop any operation which the Safety Director or his/her designated representative determines is not safe.

It is the duty of each employee to cooperate with the System Safety Department and provide it with any requested information to help in any investigation or inspection it may undertake.

———————————————————
Chief Executive Officer's Signature

Figure 3.1 Chief executive officer's statement.

with the technical information that was required for analysis that had to be performed before design decisions were made. This was evident in the Failure Mode Effects Analysis (FMEA), which examines the failure modes of components or subassemblies. Many of these failure modes correlate to hazards. Differences in approach were evident in that reliability considered the failure effect with a prediction of the occurrence probability and the identification of a need to control the failure. System safety utilizes that same information in evaluating whether failure modes are a *hazard* requiring system safety analysis. Both disciplines usually report their knowledge to common program managers and participate in design reviews.

As a total discipline, product assurance represents the united efforts of system safety, reliability, maintainability, human factors, and quality efforts. Most companies identify product assurance as the overall organization with the above disciplines as participants. Table 3.1 provides common data that is used by system safety resulting from outputs of these other disciplines. The industry trend is to develop a product assurance organization and, thus, assure a comprehensive assurance capability of the product.

The system safety responsibility in many company organizations is placed in the product assurance department. Although it is certainly an organizational method of

Table 3.1. Common safety data from product assurance disciplines

Discipline	Data Available	Safety Impact
Reliability	Failure Modes and Effects Analysis (FMEA) of parts/subassemblies	Hazards identification
Quality	Defect list of parts/products	Defects that are safety related
Maintainability	Maintenance events in procedures for repairs	Hazards in performing maintenance—addressed by O&SHA or separate Maintenance Hazard Analysis (MHA)
Human factors	Evaluation of operator and maintainer requirements to control equipment	Man–machine hazards that must be addressed by Operating Hazard Analysis (O&SHA)

establishing a close relationship among such common disciplines as reliability, maintainability, quality assurance, and human engineering, it should be ascertained that the system safety effort may take its recommendations directly to top management. The policy and procedures document issued by top management should recognize the need for clear identification of hazards and attention to their control. System safety as a part of the product assurance organization can be the key for hazard management. This involves the ability to have a well-developed organization that is (1) effective through the product assurance area, (2) exists throughout the life cycle of the projects and programs of the company, and (3) documents the knowledge gained from system safety efforts. Figure 3.2 brings together these facets of the system safety program and attempts to illustrate the need to have all aspects addressed in the total *hazard management* control. Product assurance has been a baseline title

Figure 3.2 System safety program elements.

for these disciplines, one of which is system safety. Further discussion of this organization will be presented in Chapter 4.

3.3. SYSTEM SAFETY PROGRAM PLAN (SSPP)

The system safety program plan is the most important element in implementing a system safety program after top management and the customer have made the commitment to develop a safe system. Implementation must be clearly detailed by the management document that will identify the safety tasks required to conduct the system safety program. The SSPP becomes the formal document that describes the safety tasks required to meet system safety requirements. It will outline organizational responsibilities, methods of accomplishment, milestones, depth of effort, and integration with other engineering and management activities. An SSPP, whether prepared by a contractor in response to a contractual requirement or by a company in developing its product line, does the following:

- Describes an integrated effort within the total product or system program;
- Specifies the management review process and system safety management controls during the product design and development;
- Identifies all system safety tasks and activities that will be performed during the scope of work;
- Describes the method of conducting safety analyses (both quantitative and qualitative) for the product or system during the life cycle development;
- Identifies any unusual safety activities that must be performed as a result of state of the art development or application
- Summarizes the program activities necessary to complete the requirements of each data item requiring a safety input.

Thus, the SSPP must describe in detail how the contractor (or company) intends to manage and accomplish the detailed tasks contained in the scope of work (contract task requirements). In the case of the customer, it provides a means to understand how the contractor will perform and becomes a tool for negotiation and later auditing of the contract.

3.3.1. Elements of the System Safety Program Plan

The objectives of the SSPP should be tailored to accomplishing the end objectives of an acceptably safe system design process. A brief description of the proposed system is usually given, even if it is speculative at the time of initial program plan preparation. The objective of the SSPP will be to define a systematic approach that insures the following.

- Safe design is consistent with product or system requirements in a timely, cost-effective manner.

- Hazards associated with the system are identified and evaluated, eliminated or controlled to an acceptable level throughout the life cycle.
- Minimum risk is involved in the design, materials, testing, and production of the end item.
- Supporting safety data available from other systems are considered, where appropriate.
- Retrofit actions required to improve safety are minimized through the timely inclusion of safety features during the definition and development of the system.
- Modifications do not degrade inherent safety of the system. Safety in operation and maintenance is demonstrated and proved in deployment.
- Safety in termination of the product is established with clear procedures and methods for product elimination.

In summary, the SSPP should insure that a tailored program is developed for application to the system. The items in MIL–STD–882 provide an excellent checklist for the complete safety task areas that should be considered. Each system is different and involves judgment as to the degree of safety activities necessary. The system safety engineering report summarizing the safety action of the system is one of the more important reports to be generated by this effort. Section 3.3 outlines a SSPP that would be developed for a product or system. Note that it incorporates the many elements of the MIL–STD–882. Tailored programs may eliminate those safety activities not required because of the nature of the product (such as a previously developed safe product used in a slightly different application). Fosler (1977) has shown the utility of partitioning (tailoring) program tasks for the type of program.

Table 3.2 details the tasks that would be considered in the System Safety Program Plan (SSPP). Although this list is taken from MIL–STD–882, it is complete and should be considered as an excellent example for any organization. It covers *all* aspects of the system safety tasks in each of the life cycle areas of system development. Note the software task items. Software deserves special attention because of the control that software has on hardware. The program phases are detailed into the phases specified by Section 2.1. Careful tailoring of a program to meet the required tasks for its use in the life cycle should be considered as well as the complexity of the program requirements for the items listed. A consideration in defining the use of these tasks is the applicability codes which appear in MIL–STD–882 and are shown in Table 3.2. As a result, tailoring is to assure that the proper tasks are combined for the type of program.

The separation of the task efforts into management (MGT) and engineering (ENG) is seen in the grouping of tasks in the 100, 200, and 300 series. The 100 tasks are to be used in the system safety management to define the total program. Later, the 200 series have the more technical tasks that will govern the hazard analysis efforts during the life cycle. There are several summary documents that are labeled MGT in this class, but they represent the result of the analysis. The 300 series of tasks refer to software hazard analysis.

Table 3.2. Application matrix for system program development

Task	Title	Task Type	Program Phase CONCEPT	DEF	DEV	PROD
100	System Safety Program	MGT	G	G	G	G
101	System Safety Program Plan	MGT	G	G	G	G
102	Integration/Management of Associate Contractors, Subcontractors, and AE Firms	MGT	S	S	S	S
103	System Safety Program Reviews	MGT	S	S	S	S
104	SSG/SSWG Support	MGT	G	G	G	G
105	Hazard Tracking and Risk Resolution	MGT	S	G	G	G
106	Test and Evaluation Safety	MGT	G	G	G	G
107	System Safety Progress Summary	MGT	G	G	G	G
108	Qualifications of Key System Safety Personnel	MGT	S	S	S	S
201	Preliminary Hazard List	ENG	G	S	S	N/A
202	Preliminary Hazard Analysis	ENG	G	G	G	GC
203	Subsystem Hazard Analysis	ENG	N/A	G	G	GC
204	System Hazard Analysis	ENG	N/A	G	G	GC
205	Operating and Support Hazard Analysis	ENG	S	G	G	GC
206	Occupational Health Hazard Assessment	ENG	G	G	G	GC
207	Safety Verification	ENG	S	G	G	S
208	Training	MGT	N/A	S	S	S
209	Safety Assessment	MGT	S	S	S	S
210	Safety Compliance Assessment	MGT	S	S	S	S
211	Safety Review of ECPs and Waivers	MGT	N/A	G	G	G
212	—RESERVED—	—	—	—	—	—
213	GFE/GFP System Safety Analysis	ENG	S	G	G	G
301	Software Req. Hazard Analysis	ENG	S	G	G	GC
302	Top-Level Design Hazard Analysis	ENG	S	G	G	GC
303	Detailed Design Hazard Analysis	ENG	S	G	G	GC
304	Code-Level Software Hazard Analysis	ENG	S	G	G	GC
305	Software Safety Testing	ENG	S	G	G	GC
306	Software/User Interface Analysis	ENG	S	G	G	GC
307	Software Change Hazard Analysis	ENG	S	G	G	GC

Row groups (left margin): Tasks 100–108 = MANAGEMENT TASKS; Tasks 201–213 = ANALYSIS TASKS; Tasks 301–307 = SOFTWARE TASKS.

Notes: Task Type

    ENG—System Safety Enginering
    MGT—Management

Program Phase

    CONCEPT—Conceptual
    DEF—Definition
    DEV—Development
    PROD—Production

Applicability Codes

    S—Selectively Applicable
    G—Generally Applicable
    GC—Generally Applicable to Design Changes Only
    N/A—Not Applicable

Ref: MIL-STD-882B.

In the Department of Defense reporting system, the use of a Data Item Description (DID) is the key to defining the specific reportable item or document that must be submitted. The details of what is required are specified in the DID and can be altered with the Contract Deliverable Item List (Form 1423) that allows further direction and guidance as to the use of the DID. In the case of a system safety program requiring the use of the MIL–STD–882, the following DIDs are:

System Safety Program Plan
System Safety Hazard Analysis Report
Safety Assessment Report
System Safety Engineering Report
Engineering Change Proposal System Safety Report
Waiver or Deviation System Safety Report
Occupational Health Hazard Assessment Report
System Safety Program Progress Report

3.3.1.1. Organization of the SSPP

A well-written and comprehensive SSPP is the key to a successful safety program. Also required will be proper interfaces with other systems engineering disciplines, proper use of hazard analyses, and timely completion of the system safety tasks. The most significant aspect is the *clear identification* to project management of the responsibility points for system safety.

Lines of authority for the system safety management organization must be easily identifiable by all interfacing organizations within and outside the project. Since system safety interfaces with almost every aspect of management organization, it must have the ability to influence design, provide concise identification of hazards, and have the full cooperation of company management. The importance of interfaces cannot be stressed enough; they are the breakpoint for the success of the safety effort.

An example of progressive definition of the safety criteria is observed in the development of a subway flood pumping subsystem for the use in a subway tunnel that passes below a river. In this case during the conceptual stage the definition of the need for evacuating water from the subway track area is obvious and use of a flood pump is defined. Later, as the hazard analysis develops from the PHA to the SSHA the need for two flood pumps is defined that allows redundancy and more reliable performance to assure the removal of water. Further, this may result in identification of alarms and warnings to the Central Control Subway Headquarters so that they are immediately aware of a flooding situation. Finally, as the Systems Hazard Analysis (SHA) is performed the final need for communications to the train subsystem may be defined as a critical need for this emergency situation. Thus the initial hazard has resulted in various stages of actions for hazard control and methods to ensure that this control is actuated. Table 3.3 illustrates these actions for identifying the need for continual development and review of the conceptual hazards and how

Table 3.3. Safety criteria development subway tunnel flooding

		CONCEPTUAL	AFTER PHA	AFTER SSHA	AFTER SHA
	Initial hazard identified → Flooding of subway tunnel resulting in a Category 1 (Catastrophic) hazard				
Safety criteria development →		Design a flood pump to remove water from subway tunnel when it is detected.	Add a second flood pump for assurance of capability to remove water (Redundancy)	Provide alarms and warnings to central control headquarters	Provide alarm communication system to train operator when high water level detected
Hazard severity		I	I	I	I
Hazard probability		C	D	E	E
Hazard Risk Index (HRI)		1	2	3	3
		Unacceptable	Undesirable (management decision required)	Acceptable with review by management	Acceptable with review by management

they will be controlled by design/procedural methods to then place the emergency under control.

The design knowledge should be shared in the phases of concept, definition, and development, particularly since system safety makes its maximum contribution to safety in design definition and development. Correction after an accident or incident occurs in operation will be more costly and time consuming and will present more complications than establishing a good, well-developed, safe design during the course of the product evolution.

3.3.1.2. Program Tasks, Schedules, and Milestones

A specific list of the tasks, schedules, and milestones should be stated for the system safety program. The task schedule must correlate to the company plan and contractor requirements for the overall product development. Thus, it is important to see that proper safety activities are clearly identified in the product plan and that proper design reviews are identified as milestones.

A logical development of the safety program within each portion of the life cycle should be shown in the schedule with clear milestones at specific design reviews. The number of personnel in each life cycle phase to support system safety tasks as well as the outputs of each task should be identified. A schedule such as Table 2.1 is commonly used to identify each safety activity and the milestones. By adding another column for manloading, the hours of support to accomplish the tasks can be shown.

The use of hazard analysis during the complete life cycle is seen in Figure 3.3. This figure illustrates the ability to integrate the several analysis types into outputs

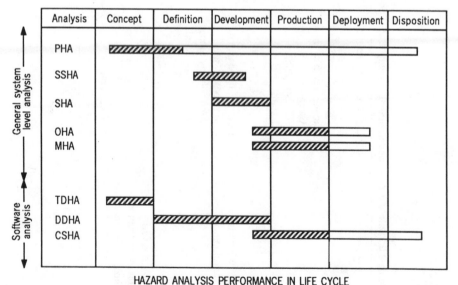

HAZARD ANALYSIS PERFORMANCE IN LIFE CYCLE

Figure 3.3 Hazard analysis performance in life cycle.

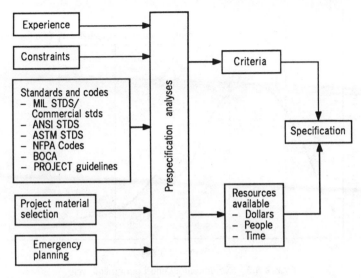

Figure 3.4 Defining safety and system criteria requirements.

that will eventually contribute to the specification for the product/system. As these analysis efforts are completed and their results show specific safety devices for the product, they will be called out in the appropriate specification for that product/ system. This is also demonstrated in Figure 3.4, which brings together information of experience, constraints, standards and codes, material selection, and emergency planning into the analyses efforts. Later, trade-offs are performed on the development of the correct criteria within the capability of resources to produce a specification that identifies controls of hazards.

Safety work load is shown in Figure 3.5, which identifies the phases of the life cycle with the magnitude of the safety effort. This vertical effort will vary with the size of the program (small, medium, or large) and implies that more effort placed in the definition and development period will result in substantial hazard control later (Petrella 1975).

3.3.1.3. Criteria
Data applicable to the system should be acquired for design criteria. Lessons learned from experience serve to provide a review of past criteria that may be of use to the evolving program. Normally the development of criteria from the concept through the development phase should be accomplished with a safety check method. Criteria should be evaluated for application through the use of the PHA and later with other analysis techniques. In this method, safety requirements evolving from identified hazards in the program can be used to validate safety criteria.

A common problem is the use of general safety requirements that are selected for safety criteria. Until the safety program is implemented, the application of safety criteria may not be possible, so a concerted effort should be made to identify a

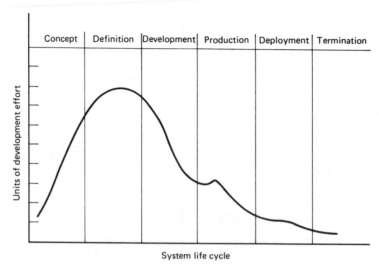

Figure 3.5 System safety effort during life cycle.

method to develop criteria. The process should be recognized as a continual updating that will not be complete until the development phase, when the characteristics of the final product or system are known.

The use of precedence in the selection of the proper hazard controls becomes a major decision point in the design development. Figure 3.6 identifies that the succession of events is to first eliminate the hazard. However this will likely be not able to be performed with most systems since the presence of energy sources is the key to their design. Thus the succeeding stages of safety controls, warnings, procedures, training, etc. become of importance. The total design effort will be built around using all of these hazard controls to reduce the impact of a total hazard defined. Table 3.3 illustrates the subway tunnel flooding that resulted in a sequence of items identified for criteria to control this hazard. Note that the hazard severity and hazard probability are marked for each stage of the items that are incorporated in the total safety criteria. Thus, as an additional pump is added the Hazard Risk Index transferred from 1 to 2. The unacceptable state of the flooding with one pump identified (Hazard Severity 1, Hazard Probability C (See definitions in Tables 1.3 and 1.4)) is an HRI 1. However in adding the second pump the HRI becomes 2, which is undesirable (Management decision required). Continuing in the incorporation of other hazard controls brings the HRI to 3 (acceptable with review by management). The information on this progression of criteria development should be documented in the PHA/SSHA/SHA and seen in the resolution column listing for action taken at decision making meetings. (See Table 3.4).

Kirkman (1980) has identified new techniques for developing safety requirements early in a program. Their usefulness is in better design to reduce the potential harm to the system. Previous NASA manned spacecraft system safety requirements showed a need for review of safety requirements resulting in continued criteria development.

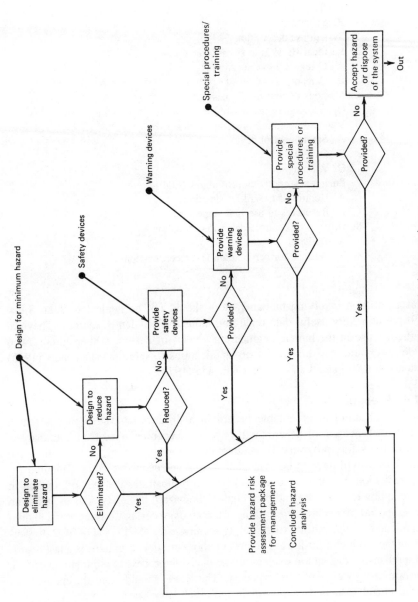

Figure 3.6 Hazard reduction precedence.

43

Table 3.4. Safety index

1. System description
   A. System
   B. Subsystems
2. System use description
3. Residual Hazards in system/subsystems
   A. Category 1—Catastrophic
   B. Category 2—Critical
4. Methods of control of hazards
5. Operator safety responsibility
6. Standard operating procedures (SOP)
   A. Safety control devices
   B. Cautions
   C. Warnings
7. Emergency operating procedures (EOP)
   A. Control of hazardous situations
   B. Resources to use in emergencies
8. Drills to be conducted in safety training
   A. Daily
   B. Periodic for certification and recertification

Design criteria involving human factors should not be neglected. Joan Surry (1969) presents very useful data regarding industrial accident resources. They are of particular use in the human engineering aspects of design. (Other authors who provide observations of human factors applications to safety are Chapanis (1965) Christensen (1981) and Van Cott and Kincade (1972).

3.3.1.4. Precedence

Designing for minimal acceptable hazard in a product or system means that a complete identification of hazards will have to be accomplished from the analysis section of the safety program. Then alternatives for eliminating or controlling the specific hazard will have to be evaluated so that an acceptable control method for hazard reduction can be agreed upon. Figure 3.6 illustrates the method applied in first attempting to eliminate and then control the hazard by safety devices, warning devices, special procedures, or training. Human engineering design criteria should be examined for applicability to the safety requirements (MIL–STD–1472D). Terminating the product if the hazard control methods are not acceptable is a last resort. A comprehensive evaluation explaining the decision process to eliminate or reduce the hazard category should be expected. The SSPP should state the methods of elimination and reduction of hazards in its initial submission.

3.3.1.5. Hazard Analyses

Hazard analyses are the heart of the system safety program. Through this process, background data from previous programs can be examined and methods to eliminate or control the hazards can be developed from the concept to the development phase.

The types of analysis that are to be performed must be stated in the SSPP and their purpose clearly defined. Since there are many types of hazard analysis, a specific attempt to understand the system and the need to perform unique types of analysis should be made.

The SSPP must describe in detail how the PHA is to be performed, what data are to be gathered, and what the result of the PHA will be. For later integration of the subsystem hazard analysis (SSHA) into the system hazard analysis (SHA), an initial description will assist in defining the effect of the hazard on the systems. Other specific analyses, such as fault tree analysis (FTA) and operating and support hazard analysis (O&SHA) should be examined, with their purpose and objective in the total safety program stated. In addition, the software analysis tasks that are key to examination of software and its interface with the total system should be examined. This includes the Top-Level Design Hazard Analysis (TDHA), Detailed Design Hazard Analysis (DDHA), and Code-Level Software Hazard Analysis (CSHA). The 300 software task requirements are further explained in Section 3.6. A total integration of anaysis efforts is illustrated in Figure 3.3. Finally, the overall method of risk analysis, incorporating information from the other analyses, should be discussed for potential application in trade-off studies and cost-benefit analysis.

3.3.1.6. Risk Assessment

Risk management is basically a decision making process consisting of evaluation and control of the probability of occurrence and the consequences of a hazardous event. This process should determine the extent and nature of preventive controls that must be applied to decrease the risk to an acceptable level. The manner in which this decision is made is critical.

Section 1.4 has defined the hazard categories and provided a hazard severity table that can be used to analyze hazards. Hazards are classified into three categories, based upon the results of design errors—human errors, environmental characteristics, and subsystem or component failures. To analyze these hazards properly, the following information should be obtained:

- Likelihood of the hazard event in the specific operating condition,
- The effect of the hazard occurrence on personnel and materials,
- The method to eliminate or control the hazard based upon cost, schedule, and product performance,
- A listing of the system safety precedence for known hazards,
- An explanation of whether the hazard could be eliminated or whether lesser methods of control should be applied,
- A definition of the acceptable level of hazard control for a specific hazard category.

The Hazard Risk Index (HRI) will become one of the primary methods of evaluation of the hazard within definitions later explained for the category of hazard and method of probability of occurrence. Risk must be addressed carefully since the total risk assessment for a hazard involves a precise understanding of the nature of a hazard and how this relates to the occurrence. Thus, a directed effort is made to clearly ensure that each hazard has a risk index associated with it and the justification of that risk. The SSPP should contain information concerning the risk assessment process and how it would be implemented in the system safety program. R. A. Hess describes very practically the usefulness of basic safety characteristics in making decisions regarding risk.

3.3.1.7. Data

The system safety process may begin with a known precedent or previous experience and knowledge. The lessons learned from a previous product or system program where system safety was applied should be used as a major source of information to benefit future programs. These data should be kept in a data bank so that both the benefits of previous safety decisions and deficiencies that have been observed will assist in developing the next program. Other manufacturing data may also be available from trade associations and common resource data banks (government, defense, and industry).

Accident and incident data should be surveyed for common types of safety-related problems that may occur in the course of the program. Engineering and procedural changes developed to control failure conditions that result in harmful events are usually available from source agencies. These agencies should be identified as sources that will be contacted in the course of the system safety program and used for the design and development of the program at hand.

Finally, there should be a clear understanding of the data that will be delivered during the course of the contract. The understanding should be stated in the SSPP, along with a brief outline of the data to be submitted. An example is hazard analysis forms, which should be included with the SSPP. The company receiving the plan will know what type of data to expect. The depth of detail of the data will highlight data differences. More detailed data may be called for. Normally a contract data requirements list (CDRL), commonly used in military contracts, is included. The purpose of the CDRL is to identify the data items to be submitted to the customer—frequency of submission, distribution, number of copies, and requirements, such as military or commercial standards. Also of importance is the data item description (DID), which is an outline of the manner in which data should be submitted. These items are generally required by commercial industry, although a data item description is not normally placed in the plan. The organization must have a single "point or contact" for system safety. This is usually the System Safety Manager or the person responsible for the implementation of the system safety program. The line of authority for the organization must clearly show the system safety manager as the "point of contact." This becomes important in large organizations with several divisions working a system safety program where one person must be the head of the organization and the prime communication contact for safety. This is discussed in Chapter 4.

Any deliverable document from the contractor to the customer becomes extremely important in subsequent litigation proceedings about a product or system. The importance level is stressed in the proper preparation of a deliverable product (Hazard Analysis, Assessment Report, Compliance Report, etc.) that offers findings in the course of the total hazard analysis effort. This has been observed continually by legal agencies who will request copies of deliverable products that may be related to the particular case that is under investigation. As a result, the system safety manager has a reponsibility in the review of work performed in the total system safety program and should carefully assure that the information in reports is fully justified. The ability to assure verification of information in reports means that references to the background data must be provided. This can take the form of standards, interviews with personnel familiar with the product, use of previous reports, testing information, and the like.

3.3.1.8. Testing and Demonstration

Testing is necessary to validate the safety of a system. Proper test planning should be done initially in the definition phase and should continue with refinements in the development and production phases. Initial qualification and prototype testing for a product that may be new or represent an innovation is required to verify the design before further development. Safety is an integral part of how that testing is conducted to insure that corrective actions incorporated in the design are verified for their effectiveness.

It will be necessary to verify, in the production phase, that a particular design meets the safety requirements specified. The SSPP should have a comprehensive listing of the requirements for safety testing, the preparation of the plans and procedures for testing, and the witnessing and evaluation of the test results. A safety test matrix to identify those particular areas where testing is required should be submitted in the SSPP. Other unknown areas can be included as the product is further defined in the course of design development.

Early planning is the key for success in assuring that the hazards that are identified are properly checked for corrective action controls. Testing milestones must be set to assure that results of the hazard analysis can be confirmed by testing. The types of tests to be considered are:

Component Testing. This is related to the actual part/component for the need to demonstrate hazard control at the lowest level. An example is a tantulum capacitor where the reverse installation of this capacitor can cause a short in a system leading to potential fire and destruction. Testing to know of the hazard control requirement and verification of the component in later installation is required.

Subassembly Testing. In the use of parts functioning together, the subassembly testing becomes most important. Knowledge of the hazards in the subassembly will have been indicated in the hazard analysis. However, testing to ensure that no new hazards occur is the purpose of this test, as well as verifying hazard controls.

Assembly Testing. With the definition of assembly, the importance of testing at the assembly level will demonstrate the assurance of controls of hazards at a higher assembly level. Some of this testing is a part of the prototype testing performed on larger groups of part combinations and modules that are identified for a specific purpose in the design of the project.

Subsystem Testing. This unified testing of the major assembly levels of a total subsystem should provide a demonstrated verification of the hazard control that is placed on the subsystem. The introduction of faults in the testing is *commonly* done to ensure that safety controls exhibit the degree of control for which they are designed.

Planning for testing and related evaluation of the testing results should include the following items:

1. Required hazard analyses are performed *prior* to the beginning of the test program so that proper test plans and procedures can be developed for safety testing.

2. Proper scheduling of analysis, evaluation, and approvals of documentation required for testing. This will introduce required safety checks in testing and ensure proper sequence of safety testing.

3. Examination of the test equipments, their instrumentation and proper installation that has been done to assure that no hazards are introduced into the test by testing equipment.

4. Identification of the environment for testing and comparison of the product to its known environmental usage. There may be specialized environmental requirements that are designated only for testing. These requirements may influence the results of testing verification of hazard control.

5. Reviewing the test reports and evaluating the follow-up action to ensure that corrective actions for hazard reductions are implemented.

3.3.1.9. Training

The company (or contractor) has a responsibility to insure that safety training for the proper use of his product is stipulated. Naturally, the qualification of one's own personnel to use the product and understand its normal operations and maintenance is of prime importance. To develop training requirements, a basic knowledge of the product and its associated hazards will be necessary. This knowledge will be useful in training personnel to avoid the occurrence of hazards later in operation and maintenance. The development of operations and maintenance manuals also requires that the personnel interfacing with the product recognize hazards and, where training is required because of the complexity and contractual requirements of the product, clearly identify them.

The focus in the SSPP is to identify the safety areas that involve specific training. Broad training requirements will ordinarily be performed by a separate group within the company, but that group is dependent upon knowledge regarding hazardous conditions of the system that it receives.

Safety training must utilize all available resource data in order to provide the latest information on the hazards of the system and then be able to train the operator of that system. Continual review of the training is required to recertify operators to ensure that they are familiar with the latest information about the system and also can demonstrate their capability to respond properly to emergency situations.

3.3.1.10. Audit Program

System safety audits should be conducted on a regular basis by an independent group either within the corporate structure or from a separate division. The purpose of the audit is to ensure that

1. The safety activities specified in the SSPP are carried out, with no problems restricting this effort.
2. Safety activities are adequate and the priority of safety work is appropriate for a tailored program.
3. Provisions of the SSPP are still applicable.
4. Deficiencies of the system safety effort are documented.

In many cases, the true purpose of the audit may not be fully understood by the safety group being inspected. The audit's tone should be constructive when improvements are suggested. The problem is to concentrate on the baseline safety plan and use this as a measuring tool against which the value of the safety program can be appraised. The progress being made will be found by discussing with the safety personnel the manner in which they are accomplishing their program and how they have overcome those problems and roadblocks. In meeting new state-of-the-art areas of development, the challenge to safety is considerable and requires an innovative approach. Again, the ability of the safety group to recognize these creative conditions of safety application is most important. Lastly, the opinion of the program management organization and other areas of the company will reflect upon the safety program and the value of its contribution to the system.

The normal audit reviews that are conducted in a program should include system safety design reviews, to assess the design and determine that the design comforms; program management reviews that are held periodically to assess the progress for all aspects of the program and the disciplines; and internal company system safety audit reviews of the progress of system safety within specific programs.

In the establishment of a safety audit, it is advisable to set up checklists of the specific items that will be reviewed. The type of Milestone checklist discussed in Table 5.2 can be used as a comprehensive checklist for a total system safety program. In addition, the normal review items of a system safety program would include the basic items listed in Table 3.5.

Notification of the system safety group well ahead of the audit is advisable to allow the group being audited to prepare documentation for review. Before starting a system safety audit, high-level management meetings with program managers should be conducted so that senior-level members of the corporation are familiar with the purpose of the audit and will take action after receiving the results.

Table 3.5. Safety audit review

The safety audit review should include a procedure and schedule for the review of the system safety program, using the SSPP as the guide. Major items to be evaluated and properly documented in this are as follows:
1. Hazard analyses that result in evaluation and control action
2. Related discipline information, such as reliability failure modes, quality defect reports, and so on
3. Safety cost benefit studies and determination of minimum acceptable level of hazards being achieved
4. Safety test plan and procedure
5. Safety equipment identified for use in the system and the criteria for selection
6. Safety criteria from hazard analyses
7. Safety action items at safety working group meetings
8. Design reviews and results of safety presentations about the design
9. Completion of safety tasks identified in the program plan

These audit items should be recorded carefully and noted in the report so that project and top-level management personnel are aware of progress. Periodic checks of subcontractors are also encouraged in order to assure the accomplishment of safety programs.

There will be times in the course of any program when an unannounced audit should be conducted. After a system safety program has been under way for a period of time, the announced audit can demonstrate the importance placed on a strong safety program and the desire of the customer to ensure that the tasks in the program plan are being completed.

NASA safety efforts have been placed in a special organization known as the Safety, Reliability, and Quality Assurance group. This is a group oriented to product assurance activities that utilize all the technical information available from each of these disciplines to assure total information exchange. Figure 3.7 illustrates the major safety interfaces and typical data flow that occurs within this group. The program activity relationships are built into the life cycle areas of the Preliminary Analysis Feasibility studies, system definition, system design and manufacture, and testing and operations. This is similar to the previous nomenclature identified in the program development for any major program.

NASA has depended upon the use of Failure Modes and Effects Analysis (FMEA) for much of the data used in the system safety hazard analysis studies. The use of this reliability technical information has proved beneficial in the accumulation of the hazardous failure modes that are then reexamined within the system safety organization for the category of the hazard and the probability of occurrence. Also, NASA has delegated much of the specific work for the development of the hazard analysis data to the subcontractors/vendors for later integration into reports providing key hazard information.

In the building construction areas, system safety is implemented with hazard analysis requirements and the use of qualified safety engineers in design as well as construction process. Comparing the life cycle development of a structure from its conceptual stage (similar to Figure 2.1) to the acceptance and use by the customer, safety is significant. Table 3.6 illustrates 10 major parts of the building process

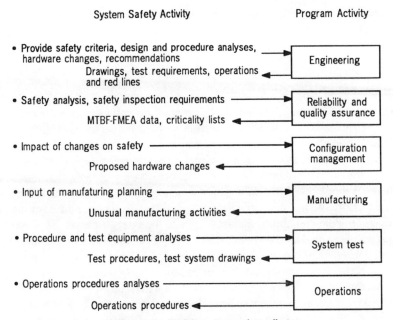

Figure 3.7   NASA system safety efforts.

Table 3.6. System safety in the construction process

Life Cycle Events	System Safety Tasks
1. Planning and land survey	Ensure that hazards at site are known
2. Architectural design	Examine safety in layout
3. Engineering design	Evaluate hazards in design
4. Approval board meeting for the project	Present safety data to board for approval
5. Construction in all phases by the construction company	Examine construction build with safety controls
6. Building checks that are made by local government	Participate in safety check
7. Building design and engineering observations made by the architect and engineer	Rexamine the build with modifications for safety controls installed
8. Completion of the "as built" drawings reflected of the actual design	Verify the changes made to safety design
9. Operational testing of the construction to perform to specifications	Demonstrate safety controls work
10. Acceptance of the project by the building inspector and finally the owner	Final acceptance tests of safety control items

where system safety has an integral contribution in assuring that hazards are first detected and controlled in facility design, with careful follow-up during the construction process.

3.4. SYSTEM INTERFACING

The SSPP should identify the interfaces with other disciplines within the company. It is very easy for each discipline to pursue its own individual objective and not share the knowledge it gains in the course of the design development process. However, the fact that system safety, by definition, interfaces with so many other aspects of the project organization necessitates that safety organization interaction be called out in the SSPP and emphasized as a major task of the organization. The integrated effort of the total organization is of key importance in ensuring that hazards identified by reliability are also known to the safety organization. Likewise, information uncovered by the system safety group in examining a design may result in the design engineering group's conducting major design changes to provide a means to reduce the effect of a particular hazard.

The overall benefit to the entire system evolution must be assessed in the planning stages of development to ensure interdisciplinary coordination. Tasks must be coordinated to ensure that each discipline can benefit from the output of the other. The result will be a more effective systems engineering and safety effort that will be much more cost effective. Outputs from each discipline, such as quality, human factors, reliability, and maintainability, must be known and coordinated for proper information exchange. The value of quality control interaction has been discussed in detail by Peters (1977) and Raheja (1978). Close coordination is a prime tool to provide visibility of potential safety problems.

An example of the system interfacing is shown in Figure 3.8 where a transit system is examined for the hazards associated with its subsystems. The train signal

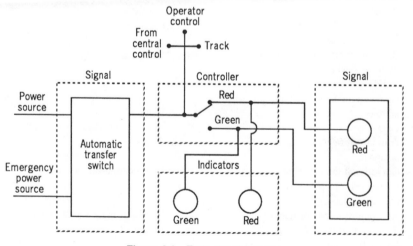

Figure 3.8   Train signal system.

system is designed to react to controls directed from the central control, operator or track to set a "green" or "red" light in the signal. Four sections of the train signal system are identified as the Signal, Controller, Indicators, and Power. The initial hazard analysis is performed on the preliminary level so that hazards for the sections are known. Then, the Subsystem Hazard Analysis (SSHA) is developed to unite the sections together and ensure that hazards that can affect the other sections of the signal system are known. Table 3.7 illustrates the Subsystem Hazard Analysis in which the automatic transfer switch of the power section has identified hazards in the failure to shift power sources when required. The hazard description states the effect of the hazard followed by the hazard effect relating to the failure of the signal to operate properly. The recommended control is later examined for methods to reduce the hazard category and probability level that is shown in the table. The importance of this subsystem is in the knowledge of the hazards associated with the sections of a subsystem and the directed attention to methods to reduce the hazard. Later, a System Hazard Analysis (SHA) can be performed considering the train signal system as a part of a total transit system where other subsystems, such as electrification, train control sytems, track, and the like, are brought together showing their interfacing and analysis results.

System safety interfaces on a regular basis with industrial safety and operational safety. The company industrial safety department functions to examine the workers and their environment. Operational safety is charged with the installation and performance of the product or system in the natural environment, in which it is designed to function. All of these safety offices require close coordination and interchange of information.

Finally, the manufacturing and transportation departments within a company will need to coordinate with the industrial and system safety groups. Examination of the product or system as it progresses through manufacturing is a key element in the process of design as well as safety. The transportation department is involved in the packaging and transport of the product or system to the customer and so must be knowledgeable about any particularly hazardous conditions of the product in the transportation mode.

3.5 HUMAN ENGINEERING

Human engineering involves the relationships between humans, procedures, machines, and the environment. Figure 3.9 illustrates the interrelationships that occur when considering these items in the ability of the system to function properly for the mission of the product or system. The system and associated equipment and facilities must be designed to promote work in the environment with proper procedures, work patterns, and personnel safety and health. Discomfort, distraction, or any other factor that degrades human performance or increases error is of major concern and must be minimized. Design is directed toward minimizing training and personnel requirements to operate and maintain the system within the limits of time, cost, and performance trade-off.

Table 3.7. Completed Signal System SSHA Form

ANALYSIS TYPE: SUBSYSTEM HAZARD ANALYSIS

SYSTEM ____ SIGNAL SYSTEM ____ PREPARED BY ____

SUBSYSTEMS ____ POWER ____ DATE ____ SHEET ____ OF ____

Item No.	Component	Function	Hazard Description	Hazard Effect	Hazard Category & Prob.	Recommended Control	Resolution
1	Automatic transfer switch	Automatically provides power from either normal or emergency source	Automatic transfer switch will *not* shift from power source to emerg. power when power source is lost	Signal fails	ID	Provide "vital signals" to shift all signals to red (stop) condition when no power is available	
2	Automatic transfer switch	Automatically provides power from either normal or emergency source	No power is available from *either* normal or emergency power source	Signal fails	IE	Provide backup battery power	

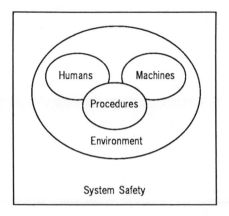

Figure 3.9. Human factors relationships between humans, procedures, machines, and environment.

Criteria must be developed to define the human limits and boundaries of operation for any product or system. This involves the knowledge of the sensory, mental, and psychomotor activities that are required to operate and maintain the system. Psychological measures that affect operation, such as a tunnel environmental for a transit rail system, must be defined in detail with all environmental parameters. The subjective responses required of the operator, maintainer, as well as the passenger (customer) must be known. Later, the frequency rate for accidents must be tracked in order to know the operational integrity of the system for human engineering design and needed improvements that should be addressed. Finally, procedures for operation of any product or system must be stated and used in the training program (standard, emergency, maintaining, etc.).

The manner in which human engineering is related to system safety is seen in the hazard impact to the person involved in any participation with the product or system. *The man–machine interface becomes the item of concern that must be addressed.* MIL–STD–882 identifies tasks that are oriented to these man–machine interface hazards. However, there is ample reason to review other hazard analysis work to ensure that known hazards are analyzed carefully for the impact on the operator, maintainer, or customer. The presence of a chemical that causes no problem to the product may have a specific problem with a person due to the toxicity level. Thus, a complete design review directed at the "person presence" with the product or system is absolutely necessary.

3.5.1 Design of the Workplace

Several factors should be considered in the design of the workplace to minimize the probability for a human error as well as hazard presence. First, equipment layout must be specifically detailed so that there are no hazards present that are unreasonable for the person interface. Knowledge of design of the equipment and its physical and functional characteristics must be documented. The complexity of the equipment should be examined for the ability of the operator to perform. All events in the life cycle of operation must be pursued (testing, producing, installing, maintaining, as well as terminating its existence).

In the Department of Defense, MIL–STD–1472D (Paragraph 5.13) is directed to the above examination and presents major areas of concern that should be considered in the overall design and operation. These include:

- Safety labels and warnings
- Facility identification
- Equipment-related hazards
- Platform
- Electrical, mechanical, or toxic hazards
- General workspace hazards
- Ionizing and nonionizing radiation

MIL–STD–1472D has been continually revised in recent years and presents basic human engineering design criteria for systems, equipment, and facilities. This standard has been applied very successfully in industry in order to promote better human engineering design and recognition of hazards involving the "man–machine" interface. Standards groups, through the encouragement of the National Bureau of Standards (now the National Institute of Standards and Technology), have recognized the human interface and the importance to design for hazard control.

3.5.2 Environment

The physical surroundings in the use of product or systems must be clearly defined. This is basically the *environment*, which has characteristic boundaries of noise, illumination, temperature, energy sources, air, humidity, shock, and vibration. During the development of the product, one of the first design definitions that will be sought is the environmental boundaries of the product and matching this with the person operating the product. Naturally the ability of the human body to withstand environments must be known. This is currently well defined in MIL–STD–1472D as well as other reference books (McCormick, Van Cott, et al.). Knowing the dimensions of the human system is necessary and specified in the anthropometry characteristics (male and female) for the recognized 5% and 95% limits of person dimensions.

All data regarding the personnel interface with the product or system should be properly recognized in human engineering data reports reviewed during the design/development time of a product. The time to change and adjust to potential hazard problems is early in the design, thereby averting operational problems that will require design changes and changes in procedures and training.

3.5.3 Causes of Errors

There are many causes of error that require attention during the design development including:

- Omission
- Commission

Figure 3.10   Failure to follow procedures.

- Nonrecognition of hazards
- Incorrect decisions
- Inadequate response
- Improper timing

Figures 3.10 to 3.12 illustrate a few human–system problems. The sequence of events in the operation of a system must be clearly defined in order to assure that

Figure 3.11   Incorrect diagnosis; misinterpretation of communications.

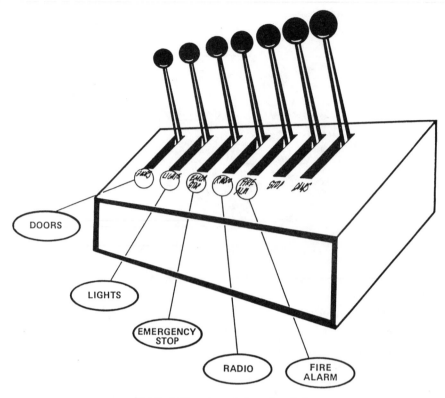

Figure 3.12 Interpretation of controls.

all aspects of event requirements are known. With this information, careful analysis of requirements for person involvement and decision making can be reviewed and assurance reached as to the maximum hazard presence. The first analysis of this aspect is normally performed in the conceptual stage of the product/system development. This is concluded by criteria that are defined for the specifications of the system. This work must recognize every aspect of activity that must be performed by the operator (or maintainer). The use of the hazard controls previously mentioned (Section 3.3.1) are enforced by warnings, training, procedures, and fail-safe devices requiring action by personnel.

THERP (Technique for Human Error Rate Prediction) is a well-developed method that has been applied effectively to deal with continuous personnel operations. The Sandia Laboratory in New Mexico (U.S. government) has conducted extensive research into boundary limits for error occurrence. These studies have dealt with methods initially used for determining the error rates in nuclear facilities.

3.5.4 Design vs. Procedural Safeguards

The human engineering design work, when properly performed in the design, will eventually lead to a preferred acceptable design. Hazard analysis will be necessary to ensure that the interface of the personnel and machine has been considered and adequately incorporated. Some of the potential problems that can occur even when the knowledge of man–machine problems are addressed by design and procedure controls are:

1. *Not following instructions.* The possibility of errors in operations of the product/system will mean that alerting designs and methods of communication should be considered for control.
2. *Doing tasks incorrectly.* If the tasks can be slipped to an incorrect method of performance easily, then efforts should be taken to ensure that checks or auditing is established.
3. *Murphy's law.* Given knowledge of the worst case situation of a potential hazardous human interface, action should be taken to develop controls to minimize this occurrence.
4. *Lack of specific knowledge.* The human relationship–interface should be reexamined carefully to ensure that all aspects of human performance are examined for potential hazards.
5. *The "It Can't Happen to Me" syndrome.* Recognize that with the use of personnel operating and using systems, the potential for believing that accidents could not happen is great. Alert messages should be used to ensure that this type of syndrome is minimized.

Human engineering is a major factor in the system safety management and engineering actions. It is noteworthy that every event needs careful examination of the actions required of the person interface with the equipment. This should result in listing hazards that can occur in the design–development life cycle. These can be controlled by design, procedures (operating, maintenance, and emergency), training, and appropriate warnings.

3.6 SOFTWARE CONSIDERATIONS

Software has developed into a major interface design to work with hardware systems. Many automated hardware units are controlled by software programs that are designed to be able to make hardware perform accurately in all operational states. The use of the manual operator to turn hardware on and off, visibly monitor, and manually operate it through its span of operation is now not needed. The use of software to perform these functions and provide monitoring status to the operator has become a normal method of design, particularly in complex systems.

Safety of the automated system is still of grave concern to the user who is responsible for the operation and maintenance of the system. In the system safety program, a separate task definition is required to perform analyses of software-controlled systems. This is briefly described in Section 1.5 and with examples of

Table 3.8. Software–hardware safety problems

Industry Category	System	Problem Situation
Space	Shuttle Challenger explosion	Booster sensors (removed) might have permitted early computer detection of leak
Rail, bus, and public transit	San Francisco BART	Power system mysteriously fails and restores itself five hours later
Commercial aviation	Air New Zealand crashes into Mt. Erebus—fatal accident	Computer course data error detected but pilots not informed
Missile, air, and naval defense	BMEWS radar system	Rising moon detected as incoming missile
Military aviation	F-16 airplane	Landing gear raised while plane on runway
Automobile	Car steering	Car with computerized steering loses control when out of gas
Electrical power	Ottawa	Power utility loses three working units due to faulty monitor

the hazard analysis methods in Part IV. The basic requirements for software safety have been addressed in DOD–STD–2167A, paragraph 4.2.3, which states:

> Safety analysis. The contractor shall perform the analysis necessary to ensure that the software requirements, design, and operating procedures minimize the potential for hazardous conditions during the operational mission. Any potentially hazardous conditions or operating procedures shall be clearly identified and documented.

Some of the software–hardware problems that have occurred within different industries are illustrated in Table 3.8. These problems demand knowledge of the software–hardware control functions. Software safety problems that can be present in the use of software with hardware systems are:

- Transfer of software type errors to the hardware resulting in improper performance
- Inability of software to respond to hardware failures
- Incomplete software programming to handle all emergency situations
- Inability of software to properly act on hazard control requirements of the hardware
- Lack of feedback from the hardware to software to verify control action
- Secondary failures emanating from software control errors

The purpose of the Task 300 series software work in MIL–STD–882, is to assure that software is thoroughly examined in the concept to design/development process to qualify it for production. This is followed by code-level software hazard analysis. Then full-scale testing with the production system is needed to verify the capability of that software to perform its tasks. Continual examination of each software change by a hazard analysis is needed to eliminate modification-introduced hazards that will then create new problems. A full acknowledgement of the need for this type of hazard analysis is essential to recognizing that a system *must* have all conditions of operational use (hardware and software) examined for hazards.

SYSTEM SAFETY MANAGEMENT ORGANIZATION

4.1. OBJECTIVES OF ORGANIZATION

The system safety management organization must be free to inquire into all areas of design, operation, and maintenance, so the system safety function must be assigned to an organizational position that will assure proper priority in obtaining information directly from all available sources. Safety engineers must have direct access to a wide variety of data and design sources to ensure that information is received promptly, and is not unconsciously changed by transfer through offices or engineering activities having potential conflicting interests. Within a company the system safety manager should be on the immediate staff of the company's program manager to insure that system safety engineering is accomplished for functions such as operational support, as well as design engineering. System safety functions best as a task group completely independent of the other supporting disciplines, such as reliability and maintainability; the group should have access to design information.

An example of the placement of safety functions within an organization is shown in Figure 4.1, which illustrates specific functions for the safety work on a detailed design level within the engineering department. Program safety and assurance activities in the program director's area and safety staff work on the corporate level should be given directly to the president or chief executive officer. Petersen (1978) has recommended, with good reason, that a safety organization be established on many levels, so that safety is examined by several different activities of the company for different purposes.

The system safety supervisor (or manager) responsible for the product design is shown in Figure 4.2. In a corresponding manner, system safety as an activity in

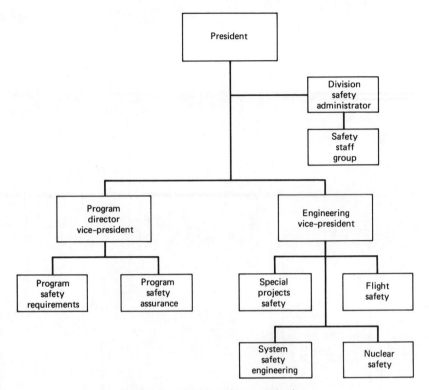

Figure 4.1   System safety organization.

engineering is also shown, but for a different purpose. Engineering task work is related to a specific requirement for assuring hazard identification, evaluation, and control in the implementation of design. However, within the engineering section there is a designated section to implement system safety criteria and design controls; for example, evolution of system safety electrical design criteria will be done by the product design group. Later, these criteria will be implemented by system safety personnel working on a specific project where safety in the electrical design is important. Also note that other disciplines—human factors, reliability analysis, and maintainability—are closely allied disciplines.

In large companies, it is evident that system safety at the corporate level is a necessity (Ferry and Weaver 1976). To have consistent application of safety in design and a proper review before the product reaches the operational stage, it is clearly valuable to have this policy implemented by a corporate director. Figure 4.3 illustrates this arrangement, with a corporate director for safety as well as a corporate system safety committee that includes members of the system safety management team in projects.

In a large organization, the use of a system safety task team (or council) is an effective means to reinforce corporate policy in each of the divisions of a company.

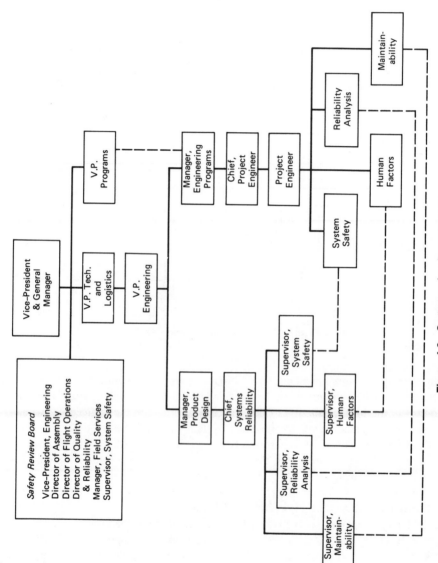

Figure 4.2 System safety in engineering.

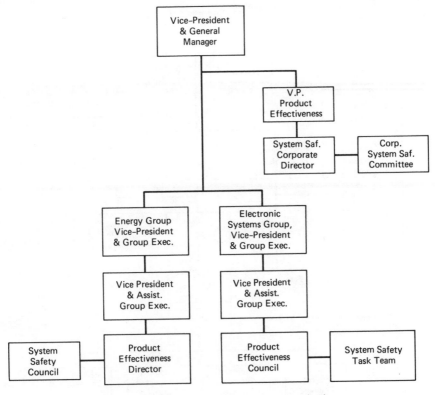

Figure 4.3 Corporate system safety organization.

Within a division, the product assurance director (or system effectiveness director) will have a staff responsible for reviewing the safety activities of the division. Figure 4.4 illustrates the relationship in which the product assurance director from the corporate level would interact with the assigned safety manager within the engineering department of the division. The system safety project engineer is responsible for the work on the project and reporting to a designated assistant chief of the engineering department. This arrangement has been found to work very effectively since it allows the corporate level staff to periodically check the progress of the system safety work and use interdivisional committees to share the safety responsibility.

The cost of a system safety program is a major factor in establishing the organization. Since the system safety program is directed at a tailored program that only selects the tasks that are required for application, cost control is mandated by the program structure. Initial analysis of task levels will be the first major item in establishing the type of system safety organization that should be developed. In large project programs, it is evident that the number of personnel will correspond to the dollar resources available as well as the complexity of the program and an initial evaluation of the types of projected hazards. In new state-of-the-art programs, there tend to

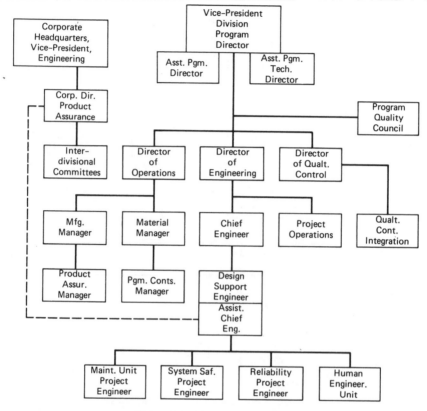

Figure 4.4   Division organization.

be larger system safety programs developed in the beginning to ensure that hazards are properly analyzed in this early concept.

A rule of thumb is that system safety programs will cost between 2% and 5% of the total program costs when they have safety requirements for design and delivery of reports to the customer. Industry reports over the last several years have also shown that the insertion of the Software Hazard Analysis has added costs that were not previously considered. The increase in software controls of hardware and the importance placed on automation and robotics for control has made it necessary to examine thoroughly the need for Software Hazard Analysis in the beginning of a program. Provisions in the System Safety Program Plan (SSPP) are made to reflect on the progress of the safety program and the need to change any of the tasks that are required (either task increases or reductions).

4.2. MANAGEMENT INTERFACE REQUIREMENTS

Safety management will require that engineering is checked regularly for the design standards that are being applied. In many cases, tailoring of programs and new products will result in many different specifications, among them, safety requirements, which must be individually certified. A significant factor in system safety effectiveness is the continual check on standards, their applicability, timeliness, and value for the particular product. Also included in the engineering interface will be the testing and demonstration activities. They should be reviewed to assure that safety tests are accomplishing their objectives.

Other interfaces include assuring that system safety is obtaining data from quality assurance regarding safety-related defects. The other assurance disciplines of reliability and maintainability will be a rich source of data regarding hazardous failure modes and failure rates. Although reliability is usually a separate function, it is important to recognize the value of information about part failures. Safety is concerned with system interactions, before and after failures, under both normal and abnormal conditions, as well as personnel error. Safety normally works closely with reliability and uses reliability input for hazard analysis.

Finally, since system safety is concerned with the total product and its potentially hazardous characteristics, it is necessary to ensure that the safety management interacts with contracts to ensure proper requirements are placed in suppliers' contracts, legal counsel, to evaluate the risks involved with the potential product, industrial safety for knowledge of environmental safety problems affecting the worker, and manufacturing for ensuring that all potential safety manufacturing problems are identified and controlled. Grimaldi and Simonds (1975) frequently point out the significance of a total safety organization that interfaces continually with all parts of the organization.

In a small organization, where the responsibility for system safety may be shared, it is important to have direct access to top management. Figure 4.5 illustrates the organization of a small company, with the system safety activity shared among the industrial safety, personnel, and fire safety offices. This organization recognizes that the work in system safety may be split among several departments. Hammer

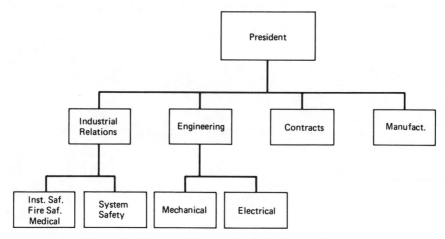

Figure 4.5 System safety organization in the small company.

(1976) recognizes shared responsibilities in the safety organization. However, sharing requires the use of other personnel in engineering to assist in accomplishing required system safety tasks. Overall responsibility and coordination of hazard analysis can be done by the engineering design personnel in the small organization, with a review by the responsible system safety person. This outside objective review is always desirable and it permits the system safety person to have direct access to top management in case difficult problems occur in the safety integrity of the product.

It is important for all concerned to understand how system safety works in a particular organization because of its involvement at one time or another with virtually every program or activity in the company. Also, it is imperative that system safety lines of authority, coordination, and administration be clearly known, so that response by the company organizational elements is indicated.

An organization is as effective as the leader of that organization desires it to be. The backing of the system safety effort must first start at the top executive level in order to achieve its goal of identification and hazards and placement of hazard control. The lines of authority that are given to the responsible system safety manager will only be as strong as the backing from the top-level management. Clear responsibility throughout the organization must be established by distinct policies and procedures.

Because of the tailoring of system safety programs to the complexity of the program and its types of hazards and dollar resources, it becomes most significant to have a clear ability to reach the managing authority (MA) for decisions. MIL-STD-882 stresses the decisions that must be made by the MA in hazard risk index (HRI) approvals and direction for hazard controls. The system safety manager must be able to reach organizationally the MA with the analyzed data and recommendations.

Qualifications of safety personnel, along with their specific education experience, certification, and training, should be identified in the company organization documents. A common deficiency in many organizations has been the use of personnel from other departments or from product assurance to man the safety effort. In many cases this has involved industrial hygiene, reliability, or quality assurance personnel taking on the system safety responsibility as an additional duty. Although it is understandable from upper management's point of view, this arrangement will not suffice when the system safety contract or product complexity requires a full-time, experienced professional.

5

SYSTEM SAFETY
CONTROL

5.1. CONTRACTOR SURVEY AND EVALUATION

The effectiveness of a system safety engineering effort is directly related to the evaluation and monitoring process that is in effect. In the initial procurement stage, the procuring agency, having selected a contractor, will want to monitor and evaluate that contractor. The contractor's system safety effort must be followed to insure compliance with contract requirements, timely completion of safety tasks, and a clear identification of hazards with effective control action implemented. Table 5.1 illustrates a checklist that can be used to conduct such a survey and evaluation.

For an industrial organization, the need to set a good example and have one's own house in order is clearly required. To conduct any type of survey, the company must be aware of the functional safety program requirements and have implemented these requirements to have achieved an initial understanding with its contractor. The contractor can be scored as meeting or not meeting requirements. In the case of a competitive bid with many contractors, it may be necessary to score so as to differentiate among them.

A source of information about government contractors is other agencies that have used the same contractor. Reputation of effort is known and methods of monitoring the contractor are documented in surveys that are conducted by assisting agencies such as Defense Logistics Agencies (DLA) or Inspector General (IG) offices. The competence of the contractor becomes known after completion of contracts and represents significant information for evaluation of contractors and their performance.

Matching the original SSPP with the performance of the contractor should provide evidence of completion and the quality of that effort.

Table 5.1. System safety review list

Contract _____ Reviewer _____ Finish _____ Contract _____

Designer _____

No.	Review Items	Review Level				Remarks
		Preliminary	Intermediate	Final	100%	
I	*Emergency storage tanks*					
	1. Are inside tanks for toxic or other than water storage vented properly to the outside?					
	2. Are underground tanks properly placed and vented?					
	3. Do tank sizes and capacities conform to specifications?					
II	*Emergency traction power trip stations*					
	1. Are trip stations located at each end of station platforms?					
	2. Are trip stations visible for ready accessibility?					
III	*Safety walks*					
	1. Are safety walks encroached upon?					
	2. Do continuous safety walks conform with design criteria?					
IV	*Emergency exits*					
	1. Are they readily accessible with no obstructions?					
	2. Are egress stairs in accordance with governing codes?					
	3. Do surface openings open into street traffic?					

The survey and evaluation that an agency can use in evaluating a contractor can be performed at any time. What is important is that every effort be made to seek information on the contractor's ability to conduct a system safety program through past experience data, present surveys, references, and other data sources that show for whom the contractor performed to the customer's requirements.

5.2. EVALUATION OF CONTRACTOR'S PROPOSALS

The evaluation of contractor proposals is accomplished by a breakdown of the proposal into the technical and management areas. Proposals are generally prepared in this manner along with a separate cost response, so that each safety cost for work to be performed in technical and management areas is detailed.

The approach to the technical area will involve careful evaluation of the analysis techniques that are proposed for conducting a comprehensive system safety engineering program. Tailoring the application of these techniques and their use for the particular project, product, or system requires judgment based on the simplicity or complexity of the project, product, or system. Tailoring tasks to achieve a comprehensive safety program is possible by partitioning, and has been shown to work effectively to meet the availability of resources (Fosler 1977). The contractor should be alerted in the request for proposal (RFP) that a survey and evaluation will be performed. Normally, the contractor will be requested in the RFP, to submit examples of hazard analysis to demonstrate his or her capability to perform in this area. The actual development of a PHA that corresponds to the proposed product/system may be suggested since this would show the ability of the contractor to perform. The items that should be covered in safety evaluation are the following.

1. *Safety Program Policy and Directives.* A detailed statement necessary to confirm contractor's commitment to enforcing a system safety program with measures to insure that the program is accomplished. The objectives should be stated clearly and responsibility assigned to the appropriate level of management. The authority for the organization and the concise statement of appropriate safety standards will form the core of the program.

2. *System Safety Program Management Plan.* A broad statement of the company management plan to show that system safety is integrated into the overall flow of development of products from concept through production and operation to termination. Included also should be identification of individuals assigned by name within the company who have direct responsibility for the accomplishment of system safety. Finally, the plan should incorporate program methods of implementation and the means to conduct surveys and evaluation of the system safety program.

3. *Safety Program Procedures.* A detailed set of program procedures that would effectively define responsibilities and methods to implement the total system safety effort in the company.

4. *Safety Design Engineering.* A statement of the safety standards that would be used in the company if the design engineering effort is required. It should be a clear, comprehensive listing of the major criteria and standards that are a part of the effort used on a regular basis in the design engineering area. Where research is required, this should be clearly shown and itemized as a task. Raouf (1980) stresses the need for proper research in this application, particularly with advanced state of the art. Finally, the capability to conduct hazard analysis must be shown.

5. *Deployment Engineering.* A statement of methods to follow up systems that are operational to insure that safety problems are recognized and appropriate action is taken. This is essentially a check on the warranty and maintenance items that are a part of the product guarantee. Appropriate follow-through with engineering changes is a necessary part of understanding how the company benefits from the knowledge of the equipment that has been in use for a period.

6. *Configuration Control.* A statement of the capability of the company to provide complete tracking of the changes introduced into a product as well as the ability to have control of the as-built drawings and specification characteristics of the product. Safety participation in change control boards (CCB) is a significant item in a good program.

7. *Safety Personnel Qualification.* A determination of the current contractor personnel capability to perform system safety work. Proper qualification, certifications, past experience, and similar information should be considered, identifying safety personnel knowledgeable to perform the contractual work. Emphasis is placed on the qualification of safety personnel performing analysis (P.E., C.S.P.) certificate. Those agencies requiring detailed analysis want to ensure that the personnel performing the analytical work are fully qualified to accomplish it. Continued emphasis on these areas is expected in the future, because it serves to insure a higher quality of safety work.

8. *Safety Data System.* The development of a data retrieval system to maintain safety data from analyses, engineering reports, and other items developed on the contract that are a formal record of accomplishment of the safety tasks is essential.

9. *Subcontractor Requirements.* The establishment of procedures to conduct surveys and evaluations of subcontractors to determine their capability to comply with system safety requirements. Examination of the ability to require subcontractors to meet the equivalent requirements for system safety should be expected.

10. *Audit System.* The development of methods and procedures to conduct a periodic audit and review of company safety efforts to insure compliance with contract requirements and schedule.

11. *Safety Training.* Procedures that identify how the safety group interacts with training to contribute methods and procedures regarding hazards and to instruct personnel in the operation or maintenance of equipment with hazards. Training contributes to operation and maintenance manuals and personnel instruction.

12. *System Safety Accident Investigations.* These should be part of the accident investigation process. Accident data should be used as a "lessons-learned" activity contributing to the future safety of developing systems.

5.3. ENGINEERING CONTROLS

Engineering controls are established through the task areas of safety analysis, requirements, design, and verification. Figure 5.2 illustrates the flow of these tasks in a satellite development project. As a result, these task areas are implemented sequentially to assure that a comprehensive system safety program is achieved. Specific contract safety requirements must be inserted as a result of understanding the product or system to be produced. These are supplemented with the efforts of a system safety working group (SSWG) composed of the principal responsible design personnel. Support activities that may be available from company-established disciplines in reliability, quality assurance, health and safety, as well as the system data history, will add to the effectiveness of engineering controls.

The results of analysis will lead to identifying specific safety requirements that can be used in the design and development of the product. Requirements are tailored for the specific application to the product and then are reviewed in the design at critical design review points to insure that they have accomplished their purpose. Along with this effort, updating of safety analysis will provide the continual history of evolution for the system. Finally, safety vertification is the means to provide testing, demonstration, audits, accident investigation, and follow up to insure that the end product has met the original safety requirements. This should result in a close-out of identified hazards that were known in the beginning of this program. An example of close-out of safety requirements is demonstrated in "Shipbuilding Safety" by Lehman (1979), where the verification of safety is required before control can be signed off as acceptable.

At any time that safety discrepancies are detected or there is doubt as to the manner to be used to control a hazard, a feedback system should be incorporated to allow review at the analysis and requirements level. This feedback system makes the greatest contribution in the design phase of engineering development of a system. Examples are seen the following types of areas:

Automobile industry (braking subsystem)	Use of brake redundancy to assure safety in ability to stop car. Alarms have been added to warn the driver and take action in several ways. Further technical advancements in antiskid computer systems have been added.
Subway transportation (ventilation subsystem)	Design of ventilation systems have seen continuous advanced changes in the automatic controls for fires. In addition, warning and monitoring subsystems have been added to ensure prompt action by the central control area for the subway system.
Chainsaws (operator chainblade subsystem)	Advancements in engineering controls are seen in the guards placed on the saw to prevent interfacing of the operator with the chain.

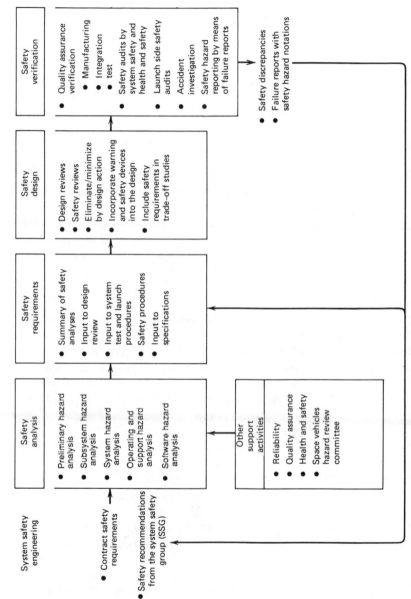

Figure 5.1. System safety engineering controls imposed on design, manufacturing, and test functions, showing action cycle taken when safety discrepancies are identified.

| Satellite systems (squib controls) | Control devices have been added to prevent operation of the saw when the hand is released from the saw control. Also, complete off conditions arise when interference within the saw blade and cutting area comes about. The initiation of many subsystem power items is dependent on squib devices. These devices are now seen as having better controls and alarms to ensure no "on" condition except when desired. The technical advancement in controls has demonstrated better operator control. |

| Power system control (emergency Action) | The emergency action in power system control has been upgraded with automated devices that will react to hazards in power downfall. Power substation controls and interaction between other substations have carefully designed methods of equalizing power outputs when problems of power generation arise. |

Thus the engineering control is a continual effort that must be observed throughout the evolution of the system. The final aspect of control will be in deployment (operational) release of the system with known hazards that cannot be further controlled or eliminated but are deemed acceptable.

5.4. MILESTONE CHECKPOINTS

One of the major methods with which to enforce a careful review of system safety is inclusion of a safety design review report at the significant checkpoints in the life cycle. This is normally established at the concept design review (CDR), preliminary design review (PDR), critical design review (CDR) and factory acceptance review (FAR) stages. Table 2.1 shows these checkpoints occurring at the end of phases of the life cycle.

There may also be interim reviews occurring in stages within each phase, such 30%, 60%, and 90% review of the definition phase leading to the final preliminary design review. These reviews very frequently occur with complex systems, where it is necessary to have several design reviews to discuss design alternatives. However, the purpose, from a system safety standpoint, is to have information communicated about the identified hazards and corrective actions. A continual updating of this information is necessary as the design matures. More technical information will allow a better decision to be made from a system safety standpoint.

A design checklist for the design review should be established so that items being designed can be properly examined and a determination made as to their

Table 5.2. Hazard identification checklist

Item	Hazard	Abbreviation	Definition	Potential Accident/Effect
1	Acceleration/shock	Accel.	Change in velocity, impact energy of vehicles, components or fluids	1. Structural deformation 2. Breakage by impact 3. Displacement of parts or piping 4. Seating or unseating valves or electrical contacts 5. Loss of fluid pressure head (cavitation) 6. Pressure surges in fluid systems 7. Detonation of shock sensitive explosives 8. Disruption of metering equipment
2	Chemical energy	Chem.	Chemical disassociation or replacement of fuels, oxidizers, explosives, organic materials or compounds	1. Fire 2. Explosion 3. Nonexplosive exothermic reaction 4. Material degradation 5. Toxic gas production 6. Corrosion fraction production 7. Swelling of organic materials
3	Contamination	Contam.	Producing or introducing contaminants to surfaces, orifices, filters, etc.	1. Clogging or blocking of components 2. Friction between moving surfaces 3. Deterioration of fluids 4. Degradation of performance sensors or operating components 5. Erosion of lines or components 6. Fracture of lines or components by fast moving large particles 7. Electrical insulation breakdown

4	Electrical energy	Elec.	System or component potential energy release or failure. Includes shock, thermal, and static	1. Electrocution 2. Involuntary personnel reaction 3. Personnel burns 4. Ignition of combustibles 5. Equipment burnout 6. Inadvertent activation of equipment or ordnance devices 7. Necessary equipment unavailable for functions or caution and warning 8. Release of holding devices 9. Interruption of communications

5	Human capability	H. Cap.	Human factors including perception, dexterity, life support, and error probability	1. Personnel injury due to:
				  a. Restricted routes
				  b. Hazardous location of
				  c. Inadequate visual/audible warnings
				2. Equipment damage by improper operation due to:
				  a. Inaccessible control location
				  b. Inadequate control/display identification
				  c. Inadequate data for decision making

6	Human hazards	H. Haz.	Conditions that could cause skin abrasions, cuts, bruises, falls, etc.	1. Personnel injury due to
				  a. Sharp edges/corners
				  b. Dangerous heights
				  c. Unguarded floor/wall openings
				  d. Limited work area

(*Table continues on p. 78.*)

Table 5.2. (*Continued*)

Item	Hazard	Abbreviation	Definition	Potential Accident/Effect
7	Interface/ Interaction	Inter.	Compatibility between systems/subsystems/ GSE/facilities/software	1. Incompatible materials reaction 2. Interfacing reactions 3. Unintended operations caused/ prevented by software
8	Kinetic energy	Kinetic	System/component linear or rotary motion	1. Linear impact 2. Disintegration of rotating components
9	Material deformation	Mat'l	Degradation of material by corrosion, aging, embrittlement, oxidation, etc.	1. Change in physical or chemical properties 2. Structural failure 3. Delamination of layered material 4. Electrical short circuiting
10	Mechanical energy	Mech.	System/component potential energy such as compressed springs	1. Personnel injury or equipment from energy release
11	Natural environment	Nat. Env.	Conditions including lightning, wind, projectiles, thermal, pressure, gravity, humidity, etc.	1. Structural damage from wind 2. Electrical discharge 3. Meteorite penetrations 4. Structural damage from space vacuum 5. Dimension changes from solar heating
12	Pressure	Press.	System/component potential energy including high, low, or changing pressure	1. Blast/fragmentation from container overpressure rupture 2. Line/hose whipping 3. Container implosion 4. System leaks 5. Heating/cooling by rapid changes 6. Aeroembolism, bends, choking, or shock

13	Radiation	Rad.	Conditions including electromagnetic, ionizing, thermal or ultraviolet radiation

1. Initiation of ordnance devices
2. Electronic equipment interference
3. Human tissue damage
4. Charring of organic materials
5. Decomposition of chlorinated hydrocarbons into toxic gases
6. Ozone or nitrogen oxide generation

14	Thermal	Therm.	System/component potential energy, including high, low or changing temperature

1. Ignition of combustibles
2. Initiation of other reactions
3. Distortion of parts
4. Expansion/contraction of solids or fluids
5. Liquid compound stratification
6. Personnel injury

15	Toxicants	Toxic	Adverse human effects of inhalants or ingesta

1. Respiratory system damage
2. Blood system damage
3. Body organ damage
4. Skin irritation or damage
5. Nervous system effects

16	Vibration/sound	Vibra.	System/component-produced energy

1. Material failure
2. Personnel fatigue or injury
3. Pressure/shock wave effects
4. Loosening of parts
5. Chattering of valves or contacts
6. Communication interference
7. Impairment or failure of displays

Table 5.3. Final acceptance safety checklist:

Question	Yes	No	Not Applicable
1. Are system safety engineering and management controls in place to ensure the system design is the safest possible consistent with mission requirements and cost effectiveness?			
2. Is management attention focusing on the identification, evaluation, and elimination or control of system hazards prior to initial operational capability?			
3. Are there any unresolved catastrophic (CAT I) and critical (CAT II) hazards associated with the system? Do those hazards present a significant safety risk?			
4. Have the actions taken to correct hazards been formally documented, including a management decision to accept the risk associated with an identified hazard?			
5. Have safety-critical issues been addressed in the Test and Evaluation Master Plan (TEMP) and evaluated during test and evaluation? Have these issues been adequately resolved as a result of the testing?			
6. Have safe operating limits been determined? Has an evaluation of mishap risk being assumed been completed and documented in a safety assessment report?			
7. Are design safety requirements, criteria, and constraints based on safety and occupational health standards and safety risk being applied and their inclusion in the system verified?			
8. Is a current system safety program in effect that meets the requirements of MIL–STD–882, System Safety Program Requirements?			
9. Is there a formal plan that integrates system safety engineering and management activities with other engineering activities to: (a) establish procedures to ensure the identification of hazards and associated risk, (b) ensure timely follow-up on identified hazards and corrective action, (c) formally designate responsibilities, including those of contractor and government system safety managers and associated organizations, (d) and allocate program funding and other needed resources?			
10. Is a trained system safety manager (principal for safety) assigned to the program manager's office?			
11. Has historical safety data (lessons learned) from previous similar acquisitions been used to prevent designing in similar safety problems?			
12. Have sufficient and appropriate system safety analyses been conducted to ensure identification of most system hazards? Are the recommended actions highlighted by these analyses complete?			
13. Have qualified personnel performed and evaluated the hazard analyses?			
14. Have the hazardous materials associated with the system, particularly for the Planned Maintenance System (PMS), been identified? Are there appropriate procedures			

Table 5.3. (*Continued*)

Question	Yes	No	Not Applicable
in place to track, store, handle, and dispose of hazardous materials/equipment? Have Material Safety Data Sheets (MSDS) been obtained from the manufacturer for all applicable material in accordance with the Code of Federal Regulations?			
15. Is a follow-on system safety effort in effect, or planned, to ensure that mission and design changes do not degrade existing levels of system safety?			

safety integrity. The checklist can take the form of a general or specific list. Where information is known about the designed item, a specific list will be most valuable. Table 5.1 presents a design checklist used for a rapid transit system. The preliminary, interim, and final review columns are provided. Additionally, comments may be listed in the right-hand column to explain noncompliance or problem areas.

A checklist that will assist in identifying hazards is shown in Table 5.2 (courtesy of USAF SAMSO, as used in SAMSO–STD–79–1). This checklist considers each type of hazard from the energy standpoint and identifies types of potential accident situations that would then be applied in reviewing a design.

A final acceptance checklist is shown in Table 5.3. This checklist is oriented to examine a complete program development to ensure that proper hazard identification, evaluation, and controls are accomplished before placing the design into an operational environment.

6

SYSTEM SAFETY IN
SYSTEM OPERATION

As has been previously stated, the objective of a continuing examination of the system for safety deficiencies throughout its service life is to bring the system to a fully acceptable state of safety, whether the user be a major company, an individual consumer, or a government. The primary effort in this examination is to insure that hazards are known, evaluated, and controlled.

During the operational phase of the system this examination continues. There is always concern for known hazards, but in addition, unknown hazards frequently emerge during the wide use of the system. Several authorities have spoken of the importance of close safety monitoring during operation (Peterson 1980, Tarrants 1980).

6.1. SYSTEM ACCEPTANCE

It has been observed that assistance for the customer should be emphasized in phasing the system into an operational status. Programs that were initiated in safety training should continue and be observed and refined in the checkout phase of the product. In the first 6 to 12 months (perhaps a guarantee period or warranty on the system), an extra effort should be made to observe the system carefully to ascertain that it is meeting its performance objectives. This monitoring includes participation by the safety management of the originating company to keep track of hazards that were identified and controlled but remain in the system. The producer will normally place a monitoring organization in charge of the phasing in of a product, to identify problem areas and take swift, efficient action to resolve them. This will minimize the cost of any warranty, as well as guard against accidents or incidents.

The methods used to minimize problems when placing a system in operation involve the close follow-up of the hazard resolution process (see Table 6.1). The importance of this process is to utilize all the known hazard data on a part/product/ system so that each of the hazards that have been identified are verified for the controls that are selected to minimize the degree of the hazard. The review technique is to review each subsystem to ensure that the controls have performed in the manner desired. Second, reviewing the total system involves examination of any unexpected surprise type of hazards that result from the total integration of the subsystems. Many times, there will be additional hazards introduced that are previously not to be a part of the design but become added through modification, testing verification, and total combination of subsystems together and performing.

Finally, the use of checklists as identified in Chapter 5 becomes a major review technique that ensures that the system has performed properly and in a safe manner.

System acceptance is also associated with a final test or demonstration that may be performed on the customer's equipment. In many cases a lengthy operational test period will be necessary to prove the reliability of the product. During this time, the producer should be carefully observing the product service, verifying those

Table 6.1. Hazard resolution process

Define the System

Define the physical and functional characteristics and understand and evaluate the people, procedures, facilities and equipment, and the environment

Identify Hazards

- Identify hazards and undesired events
- Determine the causes of hazards

Assess Hazards

- Determine severity
- Determine probability
- Decide to accept risk or eliminate/control

Resolve Hazards

- Assume risk or
- Implement corrective action
    Eliminate
    Control

Follow-up

- Monitor for effectiveness
- Monitor for unexpected hazards

failures that occur, investigating the cause of those failures, as well as proving out the maintenance capability when a failure does occur.

6.2. PARTICIPATION IN ACCIDENT INVESTIGATIONS

In spite of all design efforts made in the development of the product or system, there will still be accidents. Some of these may have been predicted; others will not have been foreseen because a different population of people use the product, and perhaps because of untested or unpredictable aspects of use or environment. Nevertheless, the safety manager must be ready to participate in accident investigations when they occur and make every effort to assure that potential hazardous situations are noted and recommendations made in the accident report to correct those safety-related problems. Peters (1962) noted that product liability has grown out of accident losses. Every effort to emphasize better safety in design will enhance the usefulness of the system and result in a decrease in accidents.

Accident investigation is a great learning experience. Careful reexamination of the system in its use to establish the cause of an accident is most important. This effort should also take place early in the design and development phase if accidents occur. Therefore, even though we discuss this task as a part of the operational safety area, it is important that accidents in any phase of the life cycle be carefully investigated.

There are several techniques utilized in an accident investigation that involve very specific examination of system safety hazard analysis data. One of the newer approaches that has proven very effective is Sequential Procedures Timed Events Plotting procedures (STEP). This method attempts to logically gather facts and time-event them in a multiplexing scheme so that at the conclusion of the investigation the root cause(s) of the accident are evident (Hendrick/Benner). Data are used in the compilation of facts, circumstances, and events that lead to the conclusion as to the causes of the accident. Associated with this final result will be the detection of any other hazards that were found in the course of examination of the data. Some of the major items to consider for safety data that also may hold potential information on hazards are:

- Statements by witnesses, transcripts, and accumulated testimony gathered in the preliminary investigation phase.
- Expert testimony by technical personnel who have examined the data of the accident and have drawn conclusions about the accident cause and its associated hazards.
- Investigators working papers that represent data used by the reviewers to draw conclusions.
- Test reports on equipments that were performed to find facts about the performance of the product/system and how it was involved in the accident. Naturally hazards may be associated with these facts of the performance.

- Program management documents that give guidance to the investigation and findings of hazards.
- Occupational Safety and Health Administration (OSHA) reports and insurance forms that have attached data regarding the accident.

The "lessons learned" from an accident can be vital in identifying specific hazards to be addressed. Also, trends in a series of accidents can point to hazard issues that need to be evaluated. Accident investigation can occur at any time in the course of the life cycle. It should not be neglected by system safety personnel.

6.3. ENGINEERING CHANGES

As a result of possible accidents or incidents with the product or system during operational use, it may be advisable to initiate an engineering change proposal (ECP). Safety management must be brought into engineering or procedural changes to consider safety aspects of changes made to the product, so that no new hazards are created as the change is incorporated (or that they are known and assessed). From the standpoint of the product, there must also be a regular examination by other groups within the company, such as material review boards (MRB), or manufacturing, and contracts, to assess the total aspect of the change and its impact. The importance of conducting a risk analysis, as discussed in Chapter 32, should be stressed. Categories of any new hazards as well as those that may be controlled further by the engineering change should be acknowledged. Hammer (1980) warns of new safety problems being introduced by engineering changes to improve safety.

The importance of safety review is the reexamination of the hazards that were originally found in the item that is being modified. First, a complete review of the hazards is made with a reevaluation of whether this new modification creates any new hazards by its introduction into the design. Second, this is followed by reviewing the hazard controls that are in place at this time to determine the effectiveness of these controls. It may be necessary to conduct additional hazard analyses based on the type of modification that is suggested by the engineering control. When this is necessary an objective independent examination must be performed to ensure a complete hazard identification, evaluation, and control.

An example of the above actions is in the space programs where the complex nature of the designed equipment is fully operated in many different environments (from preparatory launch to eventual space placement and potential return to earth). The need for control is essential as future changes must be examined as to their effect on the total design and their ability to perform under all conditions of environment. Also, the operator must be considered because of the man–machine interface being critical in the space environment. As a result, a review of OSHA is a requirement.

The use of a prototype of the modification should be considered, along with other aspects of simulation and training to verify the adequancy of the change prior to its incorporation into the product. In most cases the costs for developing the engineering change after the product or system has been produced and deployed may be much higher than the costs associated with an original design. The management of a company must constantly assess the technical, cost, and schedule commitments in light of the company objectives for products. A careful monitoring and surveillance effort during the first phase of operational deployment is mandatory. Later, a continual check on the product and reporting of any type of problem must be carried out.

6.4. TRAINING

The use of training programs that are appropriate to the users is a necessity. With many systems there will also have to be emergency training associated with handling the most critical or catastrophic hazards. As an example, consider fire in a rail transit subway when the train is in a tunnel. This situation requires detailed safety training for subway personnel on the train (as motorman or conductor), in the station, and in the train control center.

The purpose of safety training is primarily to avoid accidents/incidents. Secondary purposes are to install a positive safety attitude, impart safety knowledge and skills, build confidence, reinforce knowledge, and enforce the hazard controls. Within an organization, there are four primary types of safety training: initial, specialized, refresher, and drills. Since safety training is a control device in the hazard reduction process, it is of concern to assure that knowledge gained from earlier safety efforts is brought together in total safety training for use of the product.

Operator training is oriented toward a knowledge of the system and the ways to safely operate it. This involves the requirement to have a complete description of the system available associated with hazard controls that will properly ensure system safety. Thus, review of hazard analysis results becomes a prime training element,

Table 6.2. Safety training outline

Task	Activity
1. System description	Review hazards of system
2. Identification of equipment hazards	Hazard controls of systems (SHA)
	Hazard controls of subsystems (SSHA, PHA)
3. Identification of safety warnings	Review of warnings in controls of system
4. Operator interface with system	Review Operating and Support Hazard Analysis (OSHA)
5. How to sequentially operate the system	Review Standard Operating Procedures (SOP)
6. How to react to emergency conditions	Review Emergency Operating Procedures (EOP)
7. Review of accident background data	Lessons learned from accidents
8. Operate the system	Perform activity in exercising hazard controls
	Respond to alarms in system

followed by sequentially reviewing each phase of operator training that is required under all phases of operation of the system. Standard Operating Procedures (SOP) and Emergency Operating Procedures (EOP) are reviewed and demonstrated for the operator prior to establishing his or her qualification to operate the system. A brief outline of the major areas of safety involved in training is illustrated in Table 6.2.

Safety training requires that the user organization periodically train user personnel in how to handle emergency conditions and to assure themselves that they are using all methods that have been established. Practice may bring out changes that will enhance the system of handling both routine and emergency situations. A total understanding of these types of hazard and the means to control them is a major responsibility in qualifying a product or system for use. Grose (1971/1987) has created a priority assessment of types of hazard to facilitate control action.

6.5. USES OF DATA

Safety management should maintain communication with industry and trade association personnel to share information about their products, particularly when there are failures or accidents over long service periods. The importance of this relationship cannot be overemphasized; one company alone cannot see to it that all problems related to the product are known. A constant vigil by the industry and trade associations will provide a means to share accumulated data. Problems that might appear in one part of an industry could affect another part and, in the case of hazards, direct action is required.

An initial breakdown of the product or system in design results in the identification of the first categories of useful data from related systems, the industry, and government operation. Effort must be made to procure data that will be helpful in identifying hazards and methods of control.

For design data area, codes and standards are of major importance. These may include the American National Standards Institute (ANSI), the National Fire Protection Association (NFPA), the American Society for Testing and Materials (ASTM), building codes, and applicable military standards and specifications. There may be recommended practices and guidelines identified from the professional societies that deal in specific items used in the system.

In the case of defense projects, the Department of Defense (DOD) project offices may be major sources of data that will assist in design work. These offices are required to maintain previous records of systems and information regarding government-furnished equipment (GFE) and government-furnished materials (GFM). Similarly, each major military organization in DOD has a safety center that continually maintains the records of the use of products and systems of the DOD, in particular, the accident rates, accident causes, and recommendations made for control.

The use of Electronics Industry Association (EIA) data is an example of the method which the electronics industry has established to address electronics safety problems. Members of the industry share information by their participation on EIA committees. The Government Industry Data Electronics Program (GIDEP) SAF-ALERTS is another example of a means by which information about hazards is communicated to all participants. It provides excellent data on a minimal cost-sharing basis. Each type of industry has some form of association or group that shares information on hazards, which will benefit all and result in earlier detection and control of hazards.

In the product area, the Consumer Product Safety Commission (CPSC) has established an excellent method to acknowledge product hazards. Through the National Electronic Injury Surveillance System (NEISS), specific information from hospital emergency rooms is accumulated to identify hazards leading to personnel injuries.

Other data are available from the National Safety Council, the Department of Labor (Bureau of Labor Statistics), Occupational Safety and Health Administration, Military Safety Centers (Army, Navy, Air Force), insurance companies, industries, and so on. A thorough literature search for the product will establish numerous places where safety information can be obtained, particularly in the trade and industry associations.

Safety management plays a significant part in the data acquisition for product development. The review conducted for all reports received regarding defects, failures, and changes in the deployment phase is most important to a safety assessment of the product. During the conceptual phase (Chapter 2), previous information regarding the product and its use provide a baseline for understanding hazards that should be known in conducting an assessment of concepts for new products. The information accumulated in this effort is a keystone for the manner in which the safety program in future products and systems will be conducted.

As the sharing of information within a trade or industry can add to the knowledge of hazards, so it is with large companies and divisions, which must be aware of safety problems in products. The knowledge of safety-related problems can improve the possibility of incorporation of a hazard control or development of an engineering change. There should be a safety person on the corporate staff who has access to all divisions of the company for the purpose of data acquisition. Such a system of data acquisition and review will make substantial savings in product costs possible and will enhance the company's reputation.

 A safety data bank should be established considering the type of products and configuration control organization. The centralization of a key safety staff person to assure that hazard information is disseminated to the field is important. This staff should follow through on safety audits within the company as well as audits by the customer.

6.6. MAINTENANCE

Maintenance manuals are developed to display all required tasks during maintenance of the product. Emphasis is given to safety warnings associated with particular portions of the maintenance work and clear identification of hazards that may be encountered. Checkouts are accomplished in the development stage to assure that maintenance is performed adequately and safely prior to verifying the product for critical design review approval. Later, training is conducted with the potential user and maintenance staff to assure that they are properly trained and ready to maintain the product.

The deployment phase involves a validation that maintenance hazard areas are identified and action is taken to incorporate changes in the maintenance requirements to control hazards. A close liaison with the maintenance and service departments will be necessary so that problems in this area are brought to the attention of all concerned. It will also be beneficial to witness a maintenance task action so that the safety organization can verify the accomplishment of the maintenance task in the manner that is specified by the manual. Obviously, problems that might arise at this point can be resolved much more easily before they reach the operational phase.

When the maintenance action involves performing fault isolation and failure analysis of the product, the safety management group should participate. Frequent discussions with the maintenance personnel and verification of the types of problem that are being observed in the field are most important in gathering a broad basis of understanding of the safety problems associated with the product. This participation when maintenance is performed should be an active part of safety monitoring during the deployment (operational) phase. The analysis technique that is used in this specific area is the Maintenance Hazard Analysis (MHA). Examination is made of all the product/system operations and the use of personnel interfacing to do maintenance activities. The purpose of the MHA is to identify those hazards to personnel and equipment that may be encountered or could result in improper maintenance so that appropriate action can be taken for their elimination and control. MHA is originated prior to the first design review and is maintained current with system modification or redesign. The final analysis normally is completed prior to the start of system qualification testing. Subsequent changes to the design should have further MHA work performed to ensure that hazards in the maintenance activity are known and controlled.

Data sources used in the performance of the MHA include maintenance engineering analysis, maintenance and support plans and procedures, maintainability data, design descriptions of operating and maintenance equipment maintenance facility layout drawings, and maintenance requirement cards (MRC). The analysis approach is based on examination of each type of maintenance activity to determine if a hazard exists from its performance. Figure 6.1 illustrates a MHA.

Human engineering data is a key factor for use in the types of hazards that may be present. Section 3.5 listed several areas of concern for the man–machine interface. In this activity checklists are used with reference documentation for the type of product and its method of maintenance. The concern is personnel involvement in maintenance and attention to the potential error rate and methods of control.

ANALYSIS TYPE: MAINTENANCE HAZARD ANALYSIS

SYSTEM _____ PREVENTIVE MAINTENANCE PROCEDURE _____ PREPARED BY _____

SUBSYSTEMS _____ PUMP _____ DATE _____ SHEET _____ OF _____

Item No.	Step	Task Description	Hazard	Hazard Effect	Hazard Category & Prob.	Recommended Control	Resolution
1	3	Reverse positions of power plugs P1 and P2	Failure to perform procedure	Pump does not operate	IB	Install automatic transfer switch to switch transformer monthly or after a specified number of hours	
2	6	Test pump when procedure has been completed This step is not included in the procedure	Pump may not operate	Tunnel floods with water	ID	Add necessary steps to procedure to test operation of pump when plugs have been reversed	
3	7	Activate alarm to check that pump starts when alarm is activated This step is not included in the procedure	Pump fails to start automatically when high level alarm is received	Tunnel floods with water	ID	Write operational procedure to manually start alarm to determine that pump starts when alarm is activated	

Figure 6.1 Maintenance hazard analysis.

6.7 SYSTEM TERMINATION

The termination or elimination of a product is a safety concern. Increased emphasis today on pollution, waste products, and toxic substances contaminating our environment is an implied emphasis of the need to insure that the method of termination will not create a hazard. This problem may manifest itself in maintenance where parts of the system will be disposed of, as a result of maintenance activities. Additionally, the incorporation of engineering changes to the products will probably result in part of the product's being removed and not used again. The manner in which the removed part is taken from the system and later disposed of can be a safety matter.

Data collection methods should be defined to assist in the continual maintenance of safety information that will be analyzed at a later time. The use of operator daily logs (both automatic and manual), Management Information-Systems (MIS), personal contacts of responsibility for operations safety, normal distribution channels for information in a company, and "lesson-learned" collection points should be listed. Naturally, the information gained will be a function of the method of maintenance that is established, continual auditing, and desire of higher management to maintain data. Statistics that are gathered for periodic reporting on operating characteristics of the system are important to utilize as a source of assistance. Environmental hazards in the disposal of systems have been noted as a significant area of concern. Materials, chemical reactions and disposal methods all have to be examined for their creation of hazards. Aside from nuclear materials that are disposed of, the many waste products present in the manufacturing industry require careful identification of their hazardous effects. The life cycle of a product or system includes examination of elements in the disposal system and the outputs from using that disposal system.

Secondary effects are also of grave concern in disposal of products. The coal industry has been addressing such problems over the last several generations. These are secondary effects in the transfer of the waste products around the world by air masses. Elimination of the coal waste product at the site may solve the immediate problem in the power industry but, this in itself, creates secondary problems that must be examined. The U.S. Environmental Protection Agency (EPA) has been addressing these areas to ensure that hazard analysis efforts are performed and the total problems with waste products of any type are known and controlled.

Naturally, foresight in recognizing these problems is desirable; however, it may not always be possible, so action to review the procedures of termination are of great importance. A very active area of concern is nuclear waste disposal, which is vitally necessary at certain stages of use of nuclear systems. For this type of problem, termination represents a major task in developing a plan in which nuclear waste will be carefully handled and properly disposed of. Nuclear waste termination involves the approval of government (state and federal) agencies and, therefore, alternatives to types of disposal must be developed.

Where termination is not foreseen in the original planning, design, and development of the product, the safety manager must participate in resolution of this problem at a later time. The use of analysis methods discussed herein can be of use and should result in adequate means of control for termination.

RCISES FOR PART I   MID TERM

Introduction

1. What is the first step in the process of a disciplined approach to system safety?

2. When was the first specification developed for system safety?

3. Distinguish between system safety and system engineering.

4. Define *hazard*. What is the difference between potential and real hazard?

5. Why are accidents important to the development of a safety program?

6. What is an undesired event?

7. Define the system safety concept. Why is the life cycle important to this concept?

8. The "fly-fix-fly" concept represents a major point in the design of a system. Explain how it differs from the system safety approach.

9. What is the heart of the system safety approach?

10. What is the earliest known paper dealing with the concept of system safety?

11. Explain why the termination phase is important in the life cycle.

12. What is the objective of a system safety program? What are some of the goals involved?

13. What is the purpose of a system safety program plan?

14. What are the main parts of an accident prevention program plan?

15. Explain the four main categories of hazard.

16. Define the method of determining hazard severity.

17. In what phase does system safety make its most significant contribution?

18. Why is definition of the safety problem the most difficult portion of the safety program?

19. What are the two basic items to use in risk evaluation? LIKELY HOOD SEVERITY 15-16

20. Explain how failure modes can be hazardous events.

21. Why is system safety a major part of design of a product or system?

22. Explain how system safety is emphasized in the life cycle events of the building process.

23. How can the FMEA be used in the system safety task work?

24. Why is software analysis for hazards important to examine? Explain the relationship between software and hardware in a total system.

25. What is the purpose of the Hazard Risk Index (HRI)?   17

26. Explain how to relate probability of occurrence to quantitative values.

System Life Cycle

1. What are the five major phases of system safety? Define the sixth phase that is also associated with the life cycle.   21

2. List other names associated with the defined phases.   22

3. What are safety control points? How are they used?   24 ?

4. What group normally conducts failure analysis?   27

5. Safety criteria normally develop from the output of what safety task?   ~~DESIGN~~
CONCEPT

6. What document defines the phases and their relationships?

7. What are the three basic questions to be answered at the end of the concept   23 phase?

8. Will all the hazards about a product or system be known at the end of the concept phase? Explain.   27

9. What type of design results come from the definition phase?   27

10. What types of testing are conducted in the development phase?   27

11. Do the system safety groups normally prepare the failure modes and effects 27 analysis (FMEA)?

12. The quality control (QC) department has a major part in production. Explain how it interfaces with safety.

13. What contribution is made by safety to the training program?   27

14. The operating hazard analysis (OHA) is a particular type of analysis. Define it.

15. Can safety changes still be introduced before the deployment phase?

16. What is the purpose of the data item description (DID)?

17. Why is it necessary to examine changes in design (engineering change proposals)?

System Safety Implementation

1. Identify company program objectives for instituting a safety plan.   29

2. How is risk evaluated and used as a method for examining program objectives? 45

3. The current state of the art for safety has been said to define an acceptable level. Do you agree or disagree with this? Why?

4. How is the life cycle incorporated into the program objectives?

5. Identify the tasks in accomplishment of the program objectives for safety in a product or system development.

6. What is the most important element in implementing a system safety program?

7. How are the policies and procedures of a company important to the safety objectives?

8. Define the basic elements of the system safety program plan.

9. Is the Military Standard–882 applicable as a document for use with commercial products? Why?

10. Explain the significance of the levels of program management and its importance for system safety definition.

11. Where do system safety tasks make their most important contribution in the life cycle of a product or system?

12. An increase in system safety activity is often seen in production. Why does this occur?

13. Explain the system safety hazard reduction precedence sequence. How does it result in elimination of hazards?

14. What is the heart of the system safety program?

15. Explain the relationship that system safety tasks have to risk assessment.

16. Why is it advisable to define the safety submissions in the contract before starting the work of the contract?

17. Define the purposes of the system safety audit program. What is used as the prime document to conduct the audit?

18. Identify the major items to be evaluated in the system safety audit.

19. Identify why interfacing is a key element of the system safety program.

20. Give examples of company policies and procedures for system safety.

21. What is the process for implementing a system safety program? What is the key element in this process?

22. Does the chief executive officer (CEO) have to have a lengthy safety policy statement for his company?

23. Explain the relationship of system safety to product assurance (PA). What is the common link between the disciplines in PA?

24. Can system safety effectively operate as a part of PA?

25. What basic system safety type analysis method is used for the human factors effort?

26. Explain how the Application Matrix for System Program Development is used to tailor a system safety program plan.

27. Can a safety program be organized effectively by representatives of each department within a company to perform system safety work tasks?

28. Do the hazard analyses normally show the hazard severity and hazard probability both before and after changes initiated by the analysis results?

29. What is the most important of the identified software hazard analyses?

30. What are the most important hazard categories to stress in safety training?

31. Define the major relationships in human engineering.

32. Are human errors used in hazard analysis? If so, give some typical human error categories.

33. What is the THERP process?

34. Can real and potential hazards be defined for software?

System Safety Management Organization

1. What should the safety organization be in an independent organization? In a company?

2. What is the relationship between the corporate safety group and the direct program system safety group?

3. Differentiate the large and small organization placement of the prime safety managers.

4. System safety working groups frequently work toward common objectives noted in the design process; define these objectives.

5. Define the other specialty areas within a company that the safety organization ordinarily interfaces with.

6. What is the relationship between the industrial safety and system safety organizations? How do they differ and what are their similarities?

7. Objectives in an organization for system safety management require what major support?

8. In establishing a project organization, present the argument for placement of the system safety group as a separate group reporting to the project manager.

9. What is the boundary level (lower and upper) rate for the cost of a system safety program? Will this vary on the size of the product and its complexity?

10. Can changes to the system safety program plan be made during the life cycle of the product or system?

11. Why is it important to have a clear distinction of the managing authority (MA) in the organization? Does the MA make the final decision on the acceptance of a hazard risk?

12. Can the system safety responsibility be identified in two areas of a company structure? Where is the most effective system safety work performed in the company?

13. Where is safety auditing initiated?

System Safety Control

1. Identify the principal safety program requirements that should be evaluated in surveys.

2. What are the two major areas of a contractor's proposal? Where does system safety appear in the proposal?

3. What is the purpose of partitioning (or tailoring) for a system safety program plan?

4. Identify the basic management items in the safety controls.

5. If subcontractors are required to follow safety requirements, what type of information would be provided in the safety control section of the SSPP?

6. Safety training is usually an outgrowth of hazard control in a specific way. Identify that hazard control process leading to training.

7. Safety verification is established to check what aspect of the safety process?

8. If safety discrepancies exist, what is the procedure to insure that they are resolved?

9. Is safety control a continual process? Present some examples of how the safety control is accomplished in each portion of the life cycle.

10. Define the major checkpoints in safety milestones.

11. Design checklists have been established for a specific purpose. Discuss how to use them in safety control.

12. Why is it important to keep record documentation of all safety work?

13. Is a prime company responsible for safety of design of its subcontractor's products used in the prime company's overall system?

14. Why is system safety participation in the Configuration Control Board important?

15. Explain how a feedback system is used to check a product development.

16. Why is a safety checklist used in final acceptance work?

17. Why do system safety personnel participate in accident investigations? How is the accident information used?

System Safety in System Operation

1. Operational safety involves the safety group's performing what specific tasks?

2. Failures that occur in checkout and testing of the final product are examined also by safety personnel for what purpose?

3. Accident investigations involve safety personnel for what purpose?

4. How can the safety analysis be used in finalizing the product or system for operational use?

5. Why is it necessary to have safety review at each engineering change?

6. How does safety participate in the training programs for a product or system?

7. Identify two sources of industry information regarding hazards of products.

8. How is the safety data bank used for safety enhancement in the operational phase?

9. Many corporations now place safety responsibility on the corporate staff. Identify their responsibility during the operational phase.

10. How can safety learn about hazardous conditions during maintenance of a product?

11. What are the termination types of activity that end the operational phase of a product?

12. If a large number of accidents occur with the product or system during operation, what action should the safety manager take?

13. Explain the Hazard Resolution Process? Are risks assumed in this process?

14. Explain how the STEP process in accident investigation is used.

15. Do accident investigations occur only during the production and operation process?

16. How are safety reviews conducted?

17. What is the real purpose of safety training?

18. What hazard analyses do safety training use?

19. Is the operator of a system exposed to all hazards in a system in his or her training?

20. Why are "as-built" drawings and specifications important to the system safety personnel?

21. Do major corporations have safety centers with data?

22. Explain examples of system safety data.

23. Is the maintenance hazard analysis (MHA) normally performed in the system safety program?

24. How can secondary effects of hazards be recognized?

25. What are examples of safety audit review items?

REFERENCES AND BIBLIOGRAPHY

Bass, L., "Products Liability: Design and Manufacturing Defects," McGraw-Hill Book, 1986.

Biancardi, M.F., "Design for Product Safety," *ASSE Journal*, April 1971.

Bierlein, L.W., "Red Book on Transportation of Hazardous Materials," Van Nostrand, 1987.

Brown, D.B., *Systems Analysis and Design for Safety*, Prentice-Hall, Englewood Cliffs, N.J., 1976.

Chapanis, A., "Man-Machine Engineering," Wadsworth, 1965.

Chapanis, A., "Human Factors Engineering for Safety," *Professional Safety*, July 1980.

Childs, C.W., "Cosmetic System Safety," *Hazard Prevention*, May/June 1979.

Christensen, J.M., "The Human Element in Safety Man–Machine Systems," *Professional Safety*, March 1981.

Cranston, G.E., "NASA Contributions to System Safety and Benefit to Non-Aerospace Activities—A Survey," Cranston Research, Alexandria, Va., 1972.

Dawson, G.W., B.W. Mercer, "Hazardous Waste Management," J. Wiley & Sons, 1986.

Ferry, T. and D. Weaver, *Directions in Safety*, Charles C. Thomas, Springfield, Ill., 1976.

Ferry, T.S., "Modern Accident Investigation and Analysis," J. Wiley Co., 1988.

Fosler, E.A., "Improving the Cost Effectiveness of MIL–STD–882 Through Partitioning," *Hazard Prevention*, March/April 1977.

Fourth International System Safety Conference, Proceedings of July 1979.

Grimaldi, J. and R. Simonds, *Safety Management*, Irwin, Homewood, Ill., 1975.

Grose, V.L., "System Safety in Rapid Transit," Rail Transit Conference, April 1971.

Grose, Vernon L., "Managing Risk–Systematic Loss Prevention for Executives," Prentice Hall, 1987.

Haddon, W., Jr., "The Basic Strategies for Reducing Damage from Hazards of All Kinds," *Hazard Prevention*, September/October 1980.

Hammer, W., *Handbook of System and Product Safety*, Prentice-Hall, Englewood Cliffs, N.J., 1972.

Hammer, W., *Occupational Safety Management and Engineering*. Prentice-Hall, Englewood Cliffs, N.J., 1976.

Hammer, W., *Product Safety Management and Engineering*, Prentice-Hall, Englewood Cliffs, N.J., 1980.

Hammer, W., "Engineers Can Be Hazardous to Your Health," *Trial*, February 1981.

Harris, R.E., "About the Basics," *Hazard Prevention*, November/December 1981.

Hendrick, K., L. Benner, "Investigating Accidents with STEP," Marcel Dekker, 1987.

Henrich, H.W., D. Petersen, and N. Roose, *Industrial Accident Prevention*, 5th ed., McGraw-Hill, New York, 1980.

Hess, R.A., "System Safety and the Decision Matter," *Approach*, June 1973.

Hradesky, John L., "Productivity & Quality Improvement," McGraw-Hill Book Company, 1987.

Jerome, E.A., "Jerry," "The Genesis of Safety . . . An Historical Overview," *A Gentle Introduction to Aviation Safety*, Prentice-Hall, New York, N.Y. 1976.

Kirkman, R.A., "New Techniques for Developing Safety Requirements Early in a Program," *Hazard Prevention*, May/June 1980.

Lehman, B.J., "Shipbuilding Safety," *Hazard Prevention*, May/June 1979.

Lowrance, W.W., *Of Acceptable Risk*, William Kaufmann, Los Altos, Calif., 1976.

McCormick, Ernest J., *Human Factors in Engineering and Design*, McGraw-Hill, New York, 1988.

Malasky, S., *System Safety*, Spartan Books, New York, 1974.

Military Standard 1472D, "Human Engineering Design Criteria for Military Systems, Equipment and Facilities. U.S. Government," 1986.

Miller, C.O., "Hazard Analysis and Identification in System Safety Engineering," 1968 Annals of the Assurance Sciences, ASME, New York, 1968.

Miller, C.O., "Why System Safety?" *Technology Review*, February 1971.

Miller, C.O., "Some Principles of Aviation Accident Prevention," *Hazard Prevention*, January/February 1976.

Miller, C.O., "Safety Management Factors Germane to the Nuclear Reactor Accident at Three Mile Island, March 28, 1979," *Three Mile Island: A Report to the Commissioners and to the Public*, Vol. 2, pp. 1213–1245, May 1980.

Moriarty, B., "Life Cycle Cost of Safety Estimation," Third International System Safety Conference, Arlington, Va., October 1977.

National Aeronautics and Space Administration, *Proceedings of Government-Industry System Safety Conference*, Greenbelt, Md., May 1968.

National Aeronautics and Space Administration, *System Safety Requirements for Manned Space Flight*, January 1969.

National Aeronautics and Space Administration, *System Safety*, NHB. 1700.1(V3), March 1970.

National Aeronautics and Space Administration, *Proceedings of Government-Industry System Safety Conference*, Greenbelt, Md., May 1971.

Peters, G., *Product Liability and Safety*, Coiner Publications, Washington, D.C., 1962.

Peters, G., "How to Design a Safe Aircraft," *Hazard Prevention*, March/April 1979.

Peters, G., "Zero Product Liability—Design Defect Elimination," *Hazard Prevention*, September/October 1977.

Peters, G. and F.S. Hall, "Design for Safety," *Product Engineering*, September 13, 1965.

Petersen, D., "Safety Managment: A Human Approach," Alroy Book Co. 1988.

Peterson, D., *Safety by Objectives*, Hloray, River Vale, N.J., 1978.

Peterson, D., *Techniques of Safety Management*, 2nd ed., McGraw-Hill, New York, 1978.

Peterson, D., *Analyzing Safety Performance*, Garland STPM Press, 1980.

Petrella, A., "System Safety Engineering for Transportation," *American Road Builder*, June/July 1975.

Pope, W.D. and J.J. Cresswell, "Safety Programs Management," *Professional Safety*, ASSE, August 1965.

Raheja, D., "The Real Relationship Between Quality Control and Product Safety," *Hazard Prevention*, March/April 1978.

Raouf, A., "Safety Related Research and Its Applications," *Hazard Prevention*, January/February 1980.

Rodgers, W.P., *Introduction to System Safety Engineering*, John Wiley & Sons, New York, 1971.

Roland, H.E., "System Safety Professionals—Their Education," *Hazard Prevention*, May/June 1979.

Sax, N.I., R.J. Lewis, "Rapid Guide to Hazardous Chemicals in the Workplace," Van Nostrand, 1986.

Schultz, N., "Fire and Flammability Handbook," Van Nostrand Reinhold, 1985.

Seiden, R.M., "Product Safety Engineering for Managers: A Practical Handbook and Guide," McGraw-Hill Book Co., 1984.

Slote, L., "Handbook of Occupational Safety and Health," J. Wiley & Sons 1987.

Surry, J., *Industrial Accident Research—A Human Engineering Appraisal*, University of Toronto Press, Toronto, June 1969.

"System Safety Spreads into Industry," *Business Week*, July 17, 1971.

Tarrants, W.E., *The Measurement of Safety Performance*, Garland STPM Press, New York, 1980.

U.S. Air Force, "System Safety Management," AFSCM 727-1, January 1, 1967.

U.S. Air Force, "Introduction to System Safety for Program Managers," Air Force Inspection and Safety Center, 1974.

U.S. Air Force, "USAF System Safety Programs," AF Regulation 800-16, June 6, 1979.

U.S. Department of Defense, *U.S. Army Materiel Command Pamphlet 385-23*, "Management of System Safety," July 1969.

U.S. Department of Defense, *Weapon System Safety Guidelines Handbook*, NAV ORD OD 44942, April 1970.

U.S. Department of Defense, *AFSC Design Handbook*, DH 1-6 "System Safety," July 20, 1971.

U.S. Department of Defense, "Military Standard, System Safety Program Requirements," MIL-STD-882B, Notice 1 July 1, 1987.

U.S. Department of Defense, "Military Standard, Human Engineering Design Criteria for Military Systems, Equipment and Facilities," MIL-STD-1472D, 14 March 1989.

U.S. Department of Defense, "System Safety Engineering and Management," DOD Instruion No. 5000.36, 8 April 1986.

U.S. Nuclear Regulatory Commission, "Reactor Safety Study: An Assessment of Accident Risks in U.S. Commercial Nuclear Power Plants," (WASH 1400), (NUREG-75/014), October 1975.

Van Cott, H.P. and R.G. Kincade, "Human Engineering Guide to Equipment Design." (Revised ed.) Washington DC U.S. Government Printing Office, 1972.

Vesley, W.E., F.F. Goldberg, N.H. Roberts, and D.F. Haasl, *Fault Tree Handbook*, NUREG-0492, U.S. Nuclear Regulatory Commission, Washington, D.C., 1981.

Wadden, R.A., P.A. Scheff, "Engineering Design for the Control of Workplace Hazards," McGraw-Hill Book Co., 1987.

Weinstein, A.S., A.D. Twerski, H.R. Piehler, and W.A. Donater, *Products Liability and the Reasonably Safe Product*, John Wiley & Sons, 1978.

Woodson, W.E., "Human Factors Design Handbook," McGraw-Hill, New York, NY, 1981.

II

STATISTICAL METHODS

7

PROBABILITY—A SAFETY EVALUATION TOOL

Modern system safety and its cousin—risk analysis—may assess the likelihood and severity of hazards in a system, either qualitatively or quantitatively. Qualitative assessment has wide appeal because of its subjective approach. However, quantitative safety evaluation is coming into wider acceptance as practitioners become educated, computers with their numerical methods come into common use, and methods of system analysis grow in sophistication.

Risk assessment must deal with uncertainty. It would, indeed, be a strange world if we could know with certainty that an accident of a particular type in a given system was going to occur at a specified time. Statistical methods allow us to account for the inherent variability and uncertainty in making predictions of losses due to accidents in the uncertain future. It is probability that is the analyst's principal predictive descriptor.

The gambler, the manager, the engineer, the scientist, the person on the street must all grapple with uncertainty when making decisions for their own good and that of the organization for which they work. The weather forecaster makes predictions with the aid of probability. Probability underlies the odds quoted for a sporting event. Even a description of the future course of an illness is couched in such terms. Probability theory and statistical methods do not tell the analyst precisely what the decision should be. It does show the degree of uncertainty in deciding on a particular course. Such uncertainty may be easily translated into risk.

Therefore, it is appropriate for a treatise on system safety engineering to provide some elementary background on probability theory and methods. The reader who wishes to delve more deeply into the subject by examining the many facets of the field and studying the underlying proofs of such methods can examine any of a number of excellent works in the field. The thrust of this brief discussion is to enable the reader to *use* probability in system safety. Thus, a practical, applied approach will be taken.

There are many statistical methods that are widely used in safety. Methods such as correlation, regression, analysis of variance, and nonparametric contingency table analysis are all of great use in determining the effect of independent causation variables on a criterion of safe system performance. We will leave the pursuit of this knowledge to the reader who is attempting to become a fully formed safety researcher and analyst. There is much excellent literature in those areas. This book focuses on those statistical methods most closely related to the field of system safety and its associated methods of analysis.

7.1 PROBABILITY LAWS

If one were to ask, "what is the probability of dying?" the response would undoubtedly be 1.0. However, if one were to ask what is the probability of predicting the exact year, day, hour, minute, and second of death, the response would be very slight. As the time increment (for which the prediction is required) is divided into smaller and smaller elements, the probability approaches zero. Between these two values, 1.0 and zero, we have the range of possible probability values, 1.0 being a certainty whereas zero is considered impossible.

Probability measures may be divided into two general categories: those that are estimated before the fact or before a system is allowed to function can be thought of as *a priori* probabilities. Those that are the result of an experiment or allowing a system to function while observing the results are called *a posteriori* probabilities. An example of the former would be the roll of a fair die. Such a system can be examined and found to have six sides each associated with a number one through six. Therefore, assuming all sides are equally weighted, if the die is rolled there would be a probability 1/6 of any particular number coming up.

This example leads us to definition of probability that is appropriate for this rather simple system. Other systems will require a more complex definition.

A probability of a particular event may be defined as a dimensionless ratio of the number of times a specified event occurs to the total number of trials in which this event could occur.

Thus, if we wish to determine the *a posteriori* probability of a relay failing in a particular time period, we would test a large number of relays for this period and record the number of failures. If we tested 1000 relays and had 2 failures we would have determined a failure probability of:

$$P(f) = 2/1000 = .002 \tag{7.1}$$

This, of course, is only the failure probability established by this one test. However, at least to this point, it is our best or unbiased estimate of this probability.

It would be most useful if we could examine a relay in sufficient detail and understanding to be able to determine the true probability of failure without testing the relay. If we could do so we would have determined the *a priori* probability of relay failure.

One of the objectives of system safety analysis is to be able to determine the probabilities of accident events *a priori*. If this could be done it would not be necessary to construct and operate systems to determine if their state of safety was acceptable. We could make such a judgment *before the fact* and, if the state of safety was unacceptable, make appropriate modifications in system design to bring it to an acceptable state. Although we are far from this position, it is the goal of system safety analysts to be able to determine the state of safety of a system *a priori*.

There are several probability laws that allow one to calculate system probabilities from probabilities of events that combine to system function in some logical manner. These laws may be illustrated more simply by examining the single die system.

7.1.1 Addition Law

If we wished to determine the probability of occurrence of either of two specified numbers in a single roll of a fair die we could do so with the *addition law*. This law states that one may add the probabilities of events, any one of which will satisfy a specification of system outcome, if the events are mutually exclusive. Thus, in the case of the roll of a die:

$$P(2,3) = 1/6 + 1/6 = 1/3 \tag{7.2}$$

The restriction, mutually exclusive, merely means that if one of the events occurs, such as a 2, the other may not. Clearly, in the case of the roll of a single die, we have mutual exclusivity. If we are concerned with a system in which the failure of an electrical relay either open or closed would cause a system fault, and the probability of failure of the relay open is .1 and failure closed is .2, then the probability of this system fault is:

$$P(f) = P(\text{open, closed}) = .1 + .2 = .3 \tag{7.3}$$

The events are mutually exclusive and thus the probability of the occurrence of the fault is quite accurately calculated by the addition law.

7.1.2 Multiplication Law

If we wished to determine the probability of occurrence of two specified numbers in two rolls of a fair die, we would do so by means of the *multiplication law*. This law states that one can multiply the probabilities of two or more more events that will satisfy the specification of system outcome, to determine system function, if the events are independent. Thus, in the case of the roll of a fair die twice:

$$P(2 - 3) = 1/6 \times 1/6 = 1/36 \tag{7.4}$$

The restriction of independence merely means that in the above example, the occurrence of a 2 on the first roll of the die does not influence the probability of the occurrence of 3 on the second roll. In passing, we note that if an event is independent it is not mutually exclusive and vice versa, except in a trivial case. This will be illustrated in the section on Boolean algebra. One should also that equation (7.4) specifies order. The 2 must be rolled before the 3. If we would modify the argument to allow a 2 or 3 in any order, then the addition law would also come into play.

$$P(2,3)_2 = 1/36 + 1/36 = 2/36 = 1/18 \tag{7.5}$$

If we are concerned with the propagation of a fault in a system that contains two relays functioning independent of each other and the fault would only occur if both relays failed, the probability of each failing being .1 and .2, then the probability of system fault is:

$$P(f) = .1 \times .2 = .02 \tag{7.6}$$

This type of system will be later defined as a series system. It can be seen that the probability of system function in a series system is merely the product of the probabilities of the events in the series.

7.1.3 Complementary Law

If we wished to determine the probability of occurrence of any of several numbers in one roll of a fair die, it might be more convenient to calculate such a probability with the *complementary law*. The law states that if all possible outcomes of a situation can be partitioned into two parts and the probability of one part is known, the probability of the other part may be calculated by calculating the complement of the first part. For instance if we know the probability of rolling a 2 is 1/6, but we wished to calculate the probability of rolling a 1, 3, 4, 5, or 6:

$$P(1,3,4,5,6,) = 1 - 1/6 = 1/6 + 1/6 + 1/6 + 1/6 + 1/6 = 5/6 \tag{7.7}$$

Thus, if we are concerned with the probability of any (>0) accidents in a given system under specified operating conditions and exposure period, then it would be possible to calculate this probability by the complement of the probability of having zero accidents under these conditions and exposure.

$$P(0 \text{ acc}) = .99$$

$$P(>0 \text{ acc}) = 1 - .99 = .01 \tag{7.8}$$

7.1.4 Examples

A few additional examples will illustrate the functioning of these three laws in elementary probability. Assume that one is asked to make draws from two bags of marbles, each containing marbles of two colors, red and white. Furthermore, assume that there is a requirement to predict, a priori, the probabilities of make draws of specified colors prior to the draws. Assume that bag one contains three red and seven white marbles. Bag two contains seven red and three white marbles.

1. What are the probabilities of drawing two marbles of various colors from the two bags, drawing first from bag 1 and then from bag 2?

First Draw	Second Draw	Probability
Red	Red	$.3 \times .7 = .21$
Red	White	$.3 \times .3 = .09$
White	White	$.7 \times .3 = .21$
White	Red	$.7 \times .7 = .49$

Note that since we have described the probabilities of all possible outcomes of this specified situation, the sum of the probabilities is 1.00.

Having enumerated the basic probabilities we may now ask more sophisticated questions.

2. What is the probability of drawing a different color in each of the two draws?

$$.09 + .49 = .58$$

3. What is the probability of drawing the same color in two draws?

$$.21 + .21 = .42$$

Question 3 may be answered another way using the complementary law.

$$1 - .58 = .42$$

The question of whether to replace the marble drawn prior to the next draw was avoided by drawing just once from two different bags. We will now draw twice from the two bags to see the effect of *not* replacing the marbles prior to a second draw.

	First Bag		Second Bag	
	First Draw	Second Draw	First Draw	Second Draw
Red-red	.3	.222 (2/9)	.7	.667 (6/9)
Red-white	.3	.778 (7/9)	.7	.333 (3/9)
White-white	.7	.667 (6/9)	.3	.222 (2/9)
White-red	.7	.333 (3/9)	.3	.778 (7/9)

As before, having enumerated the probability space, we may ask more complex questions.

1. What is the probability of drawing all white in four draws?

$$.7 \times .667 \times .3 \times .222 = .0311$$

2. What is the probability of drawing red-white in each pair of draws?

$$.3 \times .778 \times .7 \times .333 = .0544$$

3. What is the probability of one color in all four draws?

$$.0311 + .3 \times .222 \times .7 \times .667 = .0622$$

4. What is the probability of at least one different color being drawn in four draws?

$$1 - .0622 = .0378$$

Other questions may be asked, but these are sufficient to illustrate the operation of the probability laws.

Let us examine questions when drawing only from bag 1:

1. What is the probability of drawing red-white-red without replacement?

$$P(\text{red-white-red}) = 3/10 \times 7/9 \times 2/8 = 7/120$$

2. What is the probability of drawing red-white-red with replacement?

$$P(\text{red-white-red}) = 3/10 \times 7/10 \times 2/10 = 21/500$$

3. What is the probability of drawing exactly two red in three draws without replacement?

$$P(2 \text{ red})_3 = P(\text{red-red-white}) + P(\text{red-white-red})$$
$$+ P(\text{white-red-red}) = (3/10 \times 2/9 \times 7/8) + (3/10 \times 7/9$$
$$+ 2/8) + (7/10 \times 3/9 \times 2/8) = 21/120$$
$$= 7/40$$

Note that both the addition and multiplication were used in this solution. Note further that although each of the three terms for each of the three orders of draw appears somewhat different, each has the same value. Thus, we can compute any one order and multiply by three for the number of orders specified by the argument.

7.2   PERMUTATIONS AND COMBINATIONS

Permutations and combinations are useful concepts when attempting to understand probabalistic methods. The concept of combination will be used to illustrate the meaning of the coefficient of a binomial term.

Analysts are frequently concerned with the number of *ways* a system may function. If, in the consideration of the number of ways, their order is important, then the concept of permutation would apply. If the number of ways is unordered, the concept of combination would be appropriate.

For instance, if one wishes to arrange eight different books on a shelf that contains eight spaces, there would seem to be an extremely large number of ways that this could be done since the switching of any two books would constitute another way. The computation of the number of ways may, at first, seem a most difficult problem. However, if one would consider that there are eight choices for the first space, and, having made that choice, there are only seven choices for the second space and so on to the last book, the solution becomes quite easy.

$$\text{Number of ways} = 8 \times 7 \times 6 \times 5 \times 4 \times 3 \times 2 \times 1 = 40,320 \quad (7.9)$$

This very large value is the number of permutations one may make with eight different objects considering their order.

Now if one were attempt to find the number of unordered ways that these eight books may be arranged on this shelf, the answer would be simply one. Since order is of no consequence, the switching of any number of books would not constitute another way. This number of unordered ways is called a combination.

A more difficult problem in which the disparity of the values is not so great would be to take the same eight different books and ask how many ordered ways may them be arranged on a shelf that has only three spaces. The solution would follow a similar pattern.

$$\text{Number of permutations} = 8 \times 7 \times 6 = 336 \quad (7.10)$$

The number of combinations or unordered ways that these books may be arranged can be found by dividing equation (7.10) by the number of ways that pertain to order.

$$\text{Number of combinations} = \frac{336}{3 \times 2 \times 1} = 56 \qquad (7.11)$$

In passing, we note that a *factorial* is a number formed by multiplying a number by one less than the previous multiplier until 1 is reached. Thus:

$$8! = 40,320 \qquad (7.12)$$

Consider a bag containing 12 marbles. How many unordered ways may 3 marbles be drawn from this bag? This is equivalent to asking how many groups of 3 may be formed from a group of 12.

$$\frac{12}{3!(12 - 3)!} = \frac{12!}{3!9!} = \frac{12 \times 11 \times 10}{3 \times 2 \times 1} = 220 \text{ combinations} \qquad (7.13)$$

The general

$$\binom{n}{x} = \frac{n!}{x!(n - x)!} \qquad (7.14)$$

This method lends itself to probabalistic calculations that might be puzzling if done the logical way.

Consider a bag with 12 marbles of which 4 are red and 8 are white. What is the probability of drawing 2 red from the bag if 4 marbles are drawn without replacement?

$$P(2r)_4 = \frac{\binom{4}{2}\binom{8}{2}}{\binom{12}{4}} = \frac{6 \times 28}{495} = \frac{56}{165} \qquad (7.15)$$

Note that this computation proceeded by finding the total number of ways that 2 reds could be drawn from the 4 available reds, and divided that number by the number of unqualified ways 4 marbles could be drawn from the bag.

The same problem calculated by the logic of the probability laws would appear as follows:

$$P(3r)_4 = \left(\frac{4}{12} \times \frac{3}{11} \times \frac{8}{10} \times \frac{7}{9}\right)\binom{4}{2} = \frac{56}{165} \qquad (7.16)$$

A few other combination examples will suffice. Let us say that we are selecting resistors for installation in a system from a group of 10. Four resistors must be installed. We know that 3 resistors in the group of 10 are defective. What is the probability that we will select 2 defective resistors in the 4 selected for installation?

$$P(2d)_4 = \frac{\binom{3}{2}\binom{7}{2}}{\binom{10}{4}} = \frac{3}{10} \tag{7.17}$$

What is the probability of selecting at least 1 defective resistor in the group of 4 selected?

$$P(>0)_4 = 1 - \frac{\binom{3}{0}\binom{7}{4}}{\binom{10}{4}} = 1 - \frac{1}{6} = \frac{5}{6} \tag{7.18}$$

If three cards are drawn from a deck of playing cards without replacement what is the probability that they will be spades?

$$P(3 \text{ spades})_3 = \frac{\binom{13}{3}}{\binom{52}{3}} = \frac{11}{850} \tag{7.19}$$

If 13 cards are drawn from a deck of playing cards without replacement, what is the probability that all 13 will be spades?

$$P(13 \text{ spades})_{13} = \frac{\binom{13}{13}}{\binom{52}{13}} = \frac{1}{635,013,559,500} \tag{7.20}$$

Let us consider the classical problem of the probability of two people in a relatively small group having the same birthday, neglecting the year of birth. This is an amusing bet that can have a probability of greater than .5 of at least two people in the group having the same birthday with a surprisingly small number of people. Although the determination of the number of people needed to create a better than an even money bet seems difficult at first presentation, it can be done quite easily using the complementary law.

Assume that there are r people in the room. The probability that no two people have the same birthday is found as follows:

$$q = \frac{365 \times 364 \times \ldots (365 - r + 1)}{365^r} \tag{7.21}$$

where q is the probability that no two people have the same birthday. The probability in question becomes $1 - q$.

A few trials will quickly show that with 23 people in the room, there will be a probability of .507 that at least 2 persons have the same birthday. The majority of that surprisingly large probability space is the probability that exactly 2 people have the same birthday. However, the bet becomes winning because of the appreciable probability of more than 2 people having the same birthday.

8

DESCRIPTIVE DATA MEASURES

When a data item is gathered about the state of safety of a system, it will group into a pattern that may be used to describe the likelihood of a data point within a certain range of values. Such a data item may be called a random variable. It may be plotted into a frequency histogram to provide a graphical representation of its frequency.

For instance, in Table 8.1, is a set of data concerning tire tread wear. Twenty-nine samples of a tire brand were driven a fixed distance over the same road surface with the resultant distribution of tread wear.

Table 8.1.

Tire tread wear, mm	<4	4 − 6	6 − 8	8 − 10	>10
Frequency	2	5	12	7	3

If the categories of tire tread wear are plotted in a histogram, they would appear as Figure 8.1.

If one may imagine reducing the interval of tread wear toward zero and increasing the number of categories toward infinity, it is not hard to imagine a continuous distribution of tread wear that would result. Such a distribution is shown as Figure 8.2.

This form of presentation of the data on tread wear may be called a probability density function or simply a frequency distribution of tread wear. In this form, it can be seen that the area under the curve of the $f(x)$ represents the probability of any amount of tread wear given this tire, driving distance, and road surface. Such a probability must be 1.0. Furthermore, it should be apparent that the integral of

113

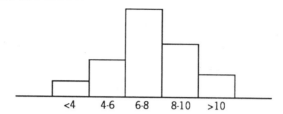

Figure 8.1   Tire treadwear histogram.

$f(x)$ between specified magnitudes of tread wear, a and b, must represent the probability of tread wear category falling between two such points.

$$P(a < \text{tread wear} < b) = \int_a^b f(x)\, dx \qquad (8.1)$$

Establishing $f(x)$ may be inexact in many cases and even if such a function can be determined with exactitude, its integration may be cumbersome. There are, however, descriptors of such a frequency distribution that can be useful in describing the underlying system from which the data are taken. There are four general categories of data descriptors: (1) measures of central tendency, (2) measures of dispersion, (3) measures of skewness, and (4) measures of kurtosis.

8.1   MEASURES OF CENTRAL TENDENCY

It is frequently useful to be able to represent an entire data set by one value. We will consider three such measures, mode, median, and mean.

The mode is that data point that appears most frequently in a data set. Thus, in any frequency distribution it would be peak of the function (Figure 8.3).

If $f(x)$ has more than one peak, the distribution may be considered by multimodal even through the peaks are not equally high. Figure 8.4 shows a bimodal distribution.

The median of a data set is the midpoint of the data if the data values are arranged in either ascending or descending order. The median of an odd number of data points would be one of the data points. Thus, the median of

3, 5, 6, 7, 9, 11, 14

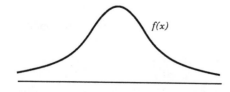

Figure 8.2   Tire treadwear distribution.

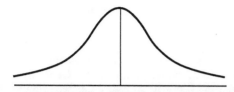

Figure 8.3 Mode of frequency distribution.

would be 7. If there are an even number of data points, then the median will be the arithmetic mean of the two central data points. Thus, the median of the following data set

$$3, 4, 5, 8, 9, 11$$

would be 6.5.

The median is coming into ever wider use as a measure of central tendency for data sets related to major public issues. This is because it is not influenced by extreme values. For instance, the typical price of a home sold in the United States is usually cited as the median. The sum of the deviations of all other data points of a set about the median will be a minimum.

The arithmetic mean is the proper name given to that measure of central tendency usually referred to as the *average*. Properly, the average may refer to any measure of central tendency. The arithmetic mean as found by adding the data items and dividing by their number. Th arithmetic mean is commonly denoted by μ. The symbol $\bar{x}$ is normally used to denote the sample mean.

$$\bar{x} = \frac{\Sigma x_i}{n} \tag{8.2}$$

where n is the number of data items.

The arithmetic mean has some special properties. Every data item is included in the computation of the value. An extreme value can exert an undue influence on the arithmetic mean. This may be either desirable or undesirable depending on the position that is being described by the statistic. Like the median, the arithmetic mean can take on values that differ from the data items. For instance, the arithmetic mean of 5, 3, 2, and 1 is 2.75.

The arithmetic mean has special significance in statistical mathematics. For instance, arithmetic mean of a data *sample* is the best estimate of the true mean of the population until additional data is gathered. Thus, it may not be altered.

Figure 8.4 Bi-modal frequency distribution.

An arithmetic mean may be weighted by multipliers of weighting factors that may be applied to some of the data items and not to others. For instance, if the arithmetic mean of a set of student grades is to be determined, but some of the grades are more important than others, the more important grades could be multiplied by weighting factors prior to adding. The sum of the weighting factors would then be applied to the divisor of the sum to find the mean. If student grades of 78, 85, 95, and 97 were to be averaged, but the grades 95 and 97 we judged twice as important as the lower two grades, the weighted arithmetic mean would be found as follows:

$$\bar{x}_w = \frac{\Sigma(W_i x_i)}{\Sigma W_i} = \frac{78 + 85 + 2(95) + 2(97)}{6} = 91.17 \qquad (8.3)$$

where W is the weighting factor.

Another measure of central tendency that is similar to the arithmetic mean, but is not so heavily weighted by extreme values, is the *geometric mean*. The geometric mean of a data set may be computed by taking the nth root of the product of the data items.

$$\bar{x}_g = \sqrt[n]{(x_1)(x_2) \dots (x_n)} \qquad (8.4)$$

The geometric of data items 3, 4, 6, 8, and 23 can be found as follows:

$$\bar{x}_g = \sqrt[5]{(3)(4)(6)(8)(23)} = 6.675 \qquad (8.5)$$

As the arithmetic mean of the data set is 8.8, it is apparent that the geometric mean was less influenced by the extreme value 23, than was the arithmetic mean.

8.2   MEASURES OF DISPERSION

Measures of dispersion or variability of a population of data items are needed along with measures of central tendency to more fully describe a data set. Such measures also assist in describing the underlying parent population of data. Measures of dispersion describe how widely the data are dispersed from the measure of central tendency. Such a measure can be thought of as setting the shape of a frequency distribution, whereas the measure of central tendency locates the distribution along the dimension axis. Measures of dispersion also assist the safety analyst in estimating how representative is the measure of central tendency. The small the dispersion the smaller is the region in which the true population mean may reside.

The measures of dispersion to be considered will be (1) range, (2) deviation, (3) variance, (4) standard deviation.

The *range* is the most direct measure of dispersion. It is merely the difference between the largest and smallest of the data items.

The *deviation* is merely the sum of the absolute magnitude of the differences of each data point from a specified point. The deviation may be taken about any point from within the data set or any other point. The deviation about zero will be the sum of the data points, considering all values positive.

The average deviation is found by dividing the total deviation by the number of data points.

$$D = \frac{\Sigma(x_i - a)}{n} \qquad (8.6)$$

where a is any defined point.

The *variance* is a more useful measure of dispersion. It is defined as the sum of the average squared deviations of the data items about their arithmetic mean. The variance is the most mathematically useful of the measures of dispersion. The variance is denoted by two symbols. The true value of variance is usually denoted by σ^2. The variance of the data sample of the parent population is commonly denoted by s^2.

The variance may be calculated as follows:

$$\sigma^2 = \frac{\Sigma(x_i - \mu)^2}{n} \qquad (8.7)$$

In the more normal circumstance, the true mean is not known. In this case, the sample variance is computed as follows:

$$s^2 = \frac{\Sigma(x_i - \bar{x})}{n - 1} \qquad (8.8)$$

It can be shown that equation (8.8) is the best estimate of the true population variance. Note that the divisor $n - 1$, increases the size of the sample variance. At some large data sample there will be little difference in the magnitude of variance whether n or $n - 1$ is used as a divisor. This value, that is normally considered the dividing point between small sample and large sample statistics, is 30.

Given the data set of 7, 8, 10, 13, and 16, with an arithmetic mean of 10.8, the sample variance may be calculated as follows:

$$s^2 = [(7 - 10.8)^2 + (8 - 10.8)^2$$

$$+ (10 - 10.8)^2 (13 - 10.8)^2$$

$$+ (16 - 10.8)^2]/4 = 13.7$$

If data points are taken from a frequency distribution of data such that a relative frequency or probability may be associated with each data item, the formula for variance becomes the following:

$$\sigma^2 = \Sigma[(x_i - E(x_i)]^2 p_i \qquad (8.9)$$

where p_i is the probability of each data item and $E(x_i)$ is the expected value of the data set.

Given the above data set with associated probabilities as follows:

$$7 - .1, \, 8 - .4, \, 10 - .2, \, 13 - .2, \, 16 - .1$$

then:

$$E(x_i) = .1(7) + .4(8) + .2(10) + .2(13) + .1(16) = 10.1$$

The population variance is them:

$$\sigma^2 = .1(7 - 10.1)^2 + .4(8 - 10.1)^2 + .2(10 - 10.1)^2$$

$$+ .2(13 - 10.1)^2 + .1(16 - 10.1)^2 = 7.89$$

Note that this is true population variance because the probabilities of each data item sum to 1.0.

The standard deviation of a data set is found by taking the square root of the variance. Thus, the standard deviation will be:

$$\sigma = \sqrt{7.89} = 2.81$$

It is recognized that this brief treatment of statistical measures will not satisfy the serious statistician. However, we include here an introduction to statistics only in sufficient depth to assist those who have had no exposure to this field. For those who wish to pursue this subject, there are a number of excellent books that examine the area more comprehensively.

9

METHODS OF SAFETY
DATA ANALYSIS

There is controversy among system safety practitioners, as to which analytical methods are properly system safety, and which reside in the more peripheral areas of safety analysis. There are those who feel that *any* analytical process concerning safety should be considered in the domain of system safety. A book that might encompass such a broad definition of system safety would then instruct in the multitude of statistical and analytical processes that could be applied to safety data and to safety problems. Such a book would need to encompass the methods found in any of a number of fine statistical texts as well as those methods found in reliability and operations research literature. While recognizing that those who attempt to take a broader view of boundaries of system safety analysis may have a valid claim, this book confines itself to the traditional and core areas of system safety analysis and a few of the more directly applicable methods from other related analytical areas.

Literature that attempts to explain methods of system safety engineering must then differentiate between the system safety peculiar methods and those techniques that are quantitative but cannot be considered in the central core of system safety analysis methods. The most closely related of these analytical methods are concerned with gathering and analyzing date to prove or disprove a safety hypothesis. This chapter lists some of these peripheral methods of data analysis that are of interest to the safety professional and explains the usefulness of each. No attempt will be made to discuss the logic of the methods nor precisely how they may be applied to safety problem areas.

9.1 CORRELATION

It is not unusual for the safety professional to be concerned about a relationship between one variable, that may possibly be a causation variable, and a criterion of safety performance such as an accident rate or severity rate. For instance, there may be concern as to whether worker training reduces the worker accident rate. The simplest method of examining such relationship, that will allow the safety professional to make estimates of the relationship between two variables, is correlation.

The *correlation coefficient* is a measure of the *linear* relationship between two variables: x as the independent or causation variable and y as the dependent or safety performance variable. The correlation coefficient is the following:

$$\sigma_{xy} = \frac{\text{cov}(x, y)}{\sigma_x \sigma_y} \tag{9.1}$$

where cov (x, y) is the covariance between x and y and σ_x, σ_y is the standard deviation of x and y.

Although equation (9.1) represents the true value of the correlation coefficient, there is a calculation formula for the sample value that can be easily followed. In equation (9.1), σ_{xy} may take values between 0.0 and 1.0 and it has a sign. A positive sign indicates direct relationship between the two variables in question. Thus, as one variable increases in some dimensional value, the other does likewise. A negative sign indicates an inverse relationship. Correlation measures the amount of this increase or decrease. Strong correlation can have either a positive or negative sign. Of course, strong correlation does not establish causation, only a statistical relationship. In the above example, it might be true that there was a strong negative correlation between training and accident rate, however on closer examination of the work situation it might be found that highly trained workers were assigned to safer jobs.

If the correlation coefficient is squared, a value called the *coefficient of determination* results. This number represents the amount of variance in one variable that may be attributed to the other variable.

If there are several variables that may influence the dependent variable a *partial* correlation coefficient may be calculated. The value will represent the amount correlation between any pair of variables holding the effects of the others constant. Of course, suitable data must be available to calculate this coefficient. There are easily followed computational methods.

9.2 REGRESSION

In correlation analysis between two variables, the analyst attempts to predict the strength and direction of the relationship between these variables. In *regression*, the analyst wishes to predict the magnitude of one variable given magnitudes of one or more independent variables. For instance, in the above example, if the analyst wished to predict the magnitude of workers' accident rate given worker training, regression would be the appropriate method to use.

Regression attempts to find a single line that best represents the relationship between the dependent variable and one or more independent variables. Linear regression assumes that this relationship is a straight line and proceeds to make the data fit a straight line. This is done by finding the means of the distributions of $E(y)$ given x. Assuming that the joint distribution of x and y is a bivariate normal distribution, the regression curve is linear, and there is one independent variable:

$$E(y/x) = \beta_1 + \beta_2 x \qquad (9.2)$$

where β_1 is y intercept of the line and β_2 is the slope of the line.

There are mathematical methods for determining the coefficients β_1 and β_2. Multiple linear regression assumes more than one independent variable. There are methods for finding the additional coefficients that will define the linear relationship $E(y/x)$ given a number of independent variables.

Nonlinear regression assumes a nonlinear relationship between several independent variables and a dependent variable. If linear regression methods are used a linear relationship will be determined regardless of the true nature of the line that best fits the data. If it is desired to predict a worker's accident rate given the worker's age, state of training, and job hazard level, multiple linear or nonlinear regression would be an appropriate method. Of course, comprehensive data sets would be needed.

9.3 ANALYSIS OF VARIANCE (ANOVA)

Analysis of variance (ANOVA) is an analytical method that examines the likelihood that independent variables have an effect on a dependent variable. It does this by examining the variances between the variables as they effect the dependent variable, compared to the variances within the variables. It is assumed that if the variances between the variables are relatively large compared to the variances within each variable, then it is more likely that the means of the variables are, indeed, different and there is a true effect of the independent variables on the dependent. The method is very robust because it allows the analyst to isolate the effects of each variable on the dependent as well as the effects of the interactions of combinations of independent variables. For the safety analyst, this method is most useful to remove the effect of one or more confounding causation variables, isolating the effect of the principal variable that is the matter of concern. For instance, age and experience are very confounding. A large value of one factor in workers, will usually mean a large value in the other. Analysis of variance will separate there effects so that may be examined individually.

One-way analysis of variance examines the effect of only one independent variable. The one-factor model is the following:

$$Y_{ij} = \mu + \alpha_j + e_{ij} \qquad (9.3)$$

where α_j is the effect of variable j and e_{ij} is the random error effect.

This model asserts that the value of observation i in a sample influenced by independent variable j, is based on the sum of three components. These are the grand mean of data items μ, an effect of the independent variable α_j, and the random error effect e_{ij}.

Two-way analysis of variance examines the effect of two independent or causation variables. The two-factor model is the following:

$$Y_{ijk} = \mu + \alpha_j + \beta_k + \gamma_{jk} + e_{ijk} \qquad (9.4)$$

In the model of equation (9.4), two factors have been added beyond that of the model of (9.3). These factors account for the additional independent variable and the interaction effect of the two variables. Of course, the random error term e_{ijk}, is due to the effects of both independent variables.

Analysis of variance relies for accuracy on two assumptions: (1) that the data samples are independently drawn from normally distributed populations and (2) that the variances of all variables are equal. The latter assumption is called homoscedasticity. Although these assumptions appear formidable, the method will yield quite good results even if they are not precisely true.

Through the use of the F distribution, which is formed by the ratio of the mean square variance between variable categories to that within each category, a significance value for the likelihood of each independent variable influencing the dependent, may be calculated.

The method can be most useful to the safety professional when attempting to isolate the effects of multiple causation variables on a measure of safety performance.

9.4 CONTINGENCY TABLES

There are a set of statistical methods that do not require assumptions about the nature of the underlying distributions such as normality, and can be used with variables that have either nominal or ordinal dimensions. Such methods are called *nonparametric methods*. One such method that is of most use in safety analysis is called *contingency table*.

If we have a data set that we believe to be under the influence of two or more categories of the independent variables, the exact probabilities of the given distribution of data items may be calculated from the *multinomial* distribution. For large numbers of data, such calculations would involve excessive work. Since the purpose of such calculations is to determine if the categories of independent variables have caused a variation in the underlying or expected distribution, there is a much simpler method of approaching such a problem.

The Pearson X^2 statistic (Karl Pearson) allows the analyst to test for independence between categorical data with a value of X^2 calculated as below:

$$X^2 = \sum \frac{(f_{oik} - f_{eik})^2}{f_{eik}} \qquad (9.5)$$

where f_{oik} is the observed frequency in cell i, k and f_{eik} is the expected frequency in cell i, k.

Assuming degrees of freedom of the number of rows minus one times the number of columns minus one, this statistic will provide a very acceptable test of independence providing, (1) each data item falls into one category, (2) the outcomes for N respective observations in the sample are independent, and (3) N is large.

9.5 SUMMARY

We have examined four well-accepted and well-documented statistical methods that will be most useful to the safety professional in analyzing safety data. These are a method of correlating one set of data with another (correlation), a method of predicting a dependent data item from values of independent items (regression), a method of separating the effects of independent variables on a dependent and determining which of the independents have a significant effect (analysis of variance), and, finally, a test for independence between sets of categorical data without an assumption of the underlying distribution (contingency table). There are many other numerical and statistical methods, particularly those that relate to reliability and maintainability of systems. This book will select the most useful of these methods with which to support the central methods of system safety analysis. For the reader who wishes to become more familiar with the peripheral safety analysis methods, references will be provided at the end of the statistical methods chapters.

10

BINOMIAL DISTRIBUTION

The binomial distribution is the first of several discrete distributions that we will examine. A discrete probability distribution is one in which there are only a finite number of values that the variable may assume. For instance, the number of children that a family may have is discrete. A family may have 1 or 2 or any number of children up to some relatively low-limiting value. A family may not have 2½ children. Likewise, the number of accidents that may occur during one hour of travel on a specified section of highway is discrete. The number may be 3 or 4, but not 3¾.

In the case of discrete distributions one may consider a *probability mass function*. Such a mass function $f(x)$, describes the probability of a random variable X, taking on exactly some value. That is $f(x) = P(X = x)$. For instance, the probability of having exactly three accidents on a given section of a highway in one hour may be calculated, given information about about the history of accidents on this section of roadway. The probability mass function has the following properties:

1. $f(x) \geq 0$ for all values of X, or $f(x)$ cannot be negative.
2. $\Sigma f(x) = 1.0$; that is the sum of the values of $f(x)$ over all values of X must equal 1.0.

Another characteristic of discrete distributions is called the *cumulative mass function*. The cumulative mass function $F(x)$, is the probability that the random variable X will be no greater than x. That is, $F(x) = P(X \leq x)$. Considering again the section of highway, the probability of having three or less accidents in one hour would be a cumulative mass function. The cumulative mass function over the range of values that X may take is 1.0.

The Bernoulli process (Jacques Bernoulli, 1654–1705) describes a series of repetitive, independent trials, the outcome of which may occur in only two ways. Such a process may be called dichotomous. Each outcome must maintain the same probability of occurrence from trial to trial. This will result if the population is essentially infinite or inexhaustible. If the process is not infinite, it will behave as such if the samples are replaced. Such a process well describes many safety situations. In such cases, an accident may occur or not on any of a number of independent exposures. Assuming that the system remains constant from exposure to exposure, or that no learning or system modification occurs, the probability of the accident event occurrence will remain constant from trial.

The binomial distribution is a very important theoretical distribution that may be applied to a Bernoulli process. It will calculate the probability of some specified number of event occurrences in n trials given the following characteristics of the process.

1. That the samples are independently and randomly selected.
2. That the probability of a particular outcome remains constant from exposure to exposure.

If the above characteristics apply to a safety situation, it is possible to calculate the probability of any specified number of accident events that may occur in a given magnitude of exposure, for instance, the probability of having one or more accidents in 10 trips of a truck.

Imagine, for a moment, that we are operating a very dangerous truck transport system, one that has a probability of an accident per trip of .1.

There only two possible outcomes of one trip of such a truck system.

$$P(\text{safe trip}) = .9$$

$$P(\text{unsafe trip}) = .1$$

Figure 10.1 illustrates.

If we were to examine the outcomes of two trips of this system and the system continues to conform to the Bernoulli process, there are four possible outcomes: an accident on both trips, an accident on the first trip and none on the second, a safe first trip and an accident on the second, or two accidents on the two trips. Figure 10.2 describes such a situation.

Neglecting the possibility of having more than one accident per trip, Figure 10.2 illustrates the calculation of the probable outcomes. It would be inconvenient to use tree diagrams to develop probabilities for greater numbers of trips. Using the

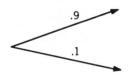

Figure 10.1   Event tree, one trip.

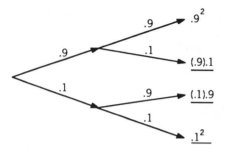

Figure 10.2   Event tree, two trips.

probability laws of Chapter 7, the probabilities of Figure 10.2 may be calculated directly.

$$P(2 \text{ acc}) = (.1)(.1) = .01$$

$$P(1 \text{ acc}) = (.1)(.9) + (.9)(.1) = .18$$

$$P(0 \text{ acc}) = (.9)(.9) = .81$$

Assuming that there are only the above possible outcomes in two trips, the probabilities thus calculated must sum to 1.0. It can be seen that they do. Then, writing our description of these events in equation form, we have the following:

$$P(2 \text{ acc}) + P(1 \text{ acc}) + P(0 \text{ acc}) = 1.0 \qquad (10.1)$$

or

$$.1^2 + 2(.1)(.9) + .9^2 = 1.00 \qquad (10.2)$$

Note that each term of equation (10.2) equals the probability of exactly some number of events. Further examination of equation (10.2) will show that it is a binomial expansion of

$$(.1 + .9)^2 = 1.00 \qquad (10.3)$$

If we assign the notation p to the action event or the probability of an accident, and q to the compliment of that event, we can rewrite equation (10.3):

$$(p + q)^2 = 1.00 \qquad (10.4)$$

In the more general case, equation (10.4) would take the following form:

$$(p + q)^n = 1.00 \qquad (10.5)$$

Expanding equation (10.5) for $n = 2$:

$$p^2 + 2pq + q^2 = 1.00 \tag{10.6}$$

If we wished to address a variety of situations and calculate probabilities for larger numbers of exposures, it would also be convenient to have a general expression for any term of the binomial series of terms. For instance, it would be far too laborious to expand equation (10.5) for $n = 10$.

Examining the three terms of the binomial series for $n = 2$, we see that each term has certain characteristics. There are coefficient and probabalistic elements of each term. The coefficient is the number of combinations of p and q in that term.

The general expression for a binomial term is then:

$$P(r/n, p) = \binom{n}{r}(p)^r(1 - p)^{n-r} \tag{10.7}$$

where r is the number of events for which the probability is being calculated.

The mean of any binomial distribution is np and variance is npq. The mean and variance is an inherent state of the system and will remain constant from term to term.

Considering again the quite unsafe truck transport system where $p = .1$, it might be of interest to examine the possible outcomes of 10 trips or exposures of this system and calculate the probabilities of the number of accidents that might occur.

$$P(0 \text{ acc}) = \binom{10}{0} (.1)^0(.9)^{10} = .3487$$

$$P(1 \text{ acc}) = \binom{10}{1} (.1)^1(.9)^9 = .3874$$

$$P(2 \text{ acc}) = \binom{10}{2} (.1)^2(.9)^8 = .1937$$

$$P(3 \text{ acc}) = \binom{10}{3} (.1)^3(.9)^7 = .0574$$

$$P(4 \text{ acc}) = \binom{10}{4} (.1)^4(.9)^6 = .0112$$

$$P(5 \text{ acc}) = \binom{10}{5} (.1)^5(.9)^5 = .0015$$

$$P(6 \text{ acc}) = \binom{10}{6} (.1)^6(.9)^4 = .0001$$

$$P(7 \text{ acc}) = \binom{10}{7} (.1)^7(.9)^3 = .0000+$$

$$P(8 \text{ acc}) = \binom{10}{8} (.1)^8(.9)^2 = .0000+$$

$$P(9 \text{ acc}) = \binom{10}{9} (.1)^9(.9)^1 = .0000+$$

$$P(10 \text{ acc}) = \binom{10}{10} (.1)^{10}(.9)^0 = .0000+$$

Having calculated wtih the binomial probabilities for any number of accidents that may occur in 10 trips or exposures, it is a simple matter to investigate the more complex questions concerning this transport system. For instance, what is the probability of 1 or more accidents?

$$P(\geq 1 \text{ acc}) = \Sigma P(1 - 10 \text{ acc}) = .6513$$

What is the probability of more than 3 accidents?

$$P(>3 \text{ acc}) = \Sigma P(4 - 10 \text{ acc}) = .0128$$

What is the probability of having more than 2, but less than 5 accidents?

$$P(>2 \text{ acc} <5) = \Sigma P(3 - 4 \text{ acc}) = .0686$$

It can be seen that the binomial distribution may be a useful tool with which to examine accident potentials and thus to judge the acceptability of the state of safety of a system.

11

MULTINOMIAL DISTRIBUTION

The logic underlying the binomial distribution can be generalized to situations that are not dichotomous but involve more than two possible outcomes. Consider the case of a transportation system where a single trip may have three possible outcomes.

No accidents	$P(0$ acc$)$	$=$	.92
Minor accident	$P($Minor acc$)$	$=$	.06
Major accident	$P($Major acc$)$	$=$	.02

Given some set of possible probabilistic outcomes in a particular safety exposure as shown above, the probability of a specified set of these outcomes may be calculated with the multinomial distribution.

For instance, what is the probability that in 15 trips of a truck system in which there exists the above potentials for safe and unsafe trips, there will be exactly 12 safe trips, 2 minor accident trips, and 1 major accident. The multinomial distribution will perform this calculation as follows:

$$P(12,\ 2,\ 1) \ = \ \frac{15!}{12!\ 2!\ 1!}\ (.92)^{12}(.06)^{2}(.02)^{1} \ = \ .0361$$

The probability is quite small, but this is to be expected considering that there are quite a large number of combinations of these three types of outcomes in 15 trips all of which must sum to 1.00.

Notice that the first term of the multinomial distribution is a fraction that represents the number of combinations of outcomes in this term.

The general expression for the multinomial distribution is as follows:

$$P(r_1, r_2, \cdots r_k/n, p_1, p_2, \cdots p_k)$$

$$= \frac{n!}{(r_1!)(r_2!), \cdots (r_k)} (p_1)^{r_1} (p_2)^{r_2} \cdots (p_k)^{rk} \quad (11.1)$$

where $r_1 + r_2 + \cdots r_k = n$, and $p_1 + p_2 \cdots p_k = 1.00$.

12

HYPERGEOMETRIC DISTRIBUTION

We have previously examined situations to which the hypergeometric probability distribution was applied, but this distribution should be examined explicitly. The distribution is similar to the multinomial, but a slight change in the process of sampling leads to a violation of the assumptions of the multinomial (stationarity and independence for p). These violations will occur when sampling from a finite population without replacement. The binomial and multinomial distributions assume that either sampling is done from an infinite population (the number of defective parts produced in a production line) or that replacement occurs.

If we are sampling from a finite population without replacement, the probability of finding a specified characteristic in the sample drawn must be altered with each sample. The probability of drawing a set with specified characteristics from the parent population may be reasoned from Section 7.2. The numerator of the probability fraction will contain the product of the number of ways each element of the set may be drawn. The denominator will be the number of ways the total sample may be drawn without restriction.

For instance, if we wish to sample from a box of 20 parts known to contain 12 parts of type I, 5 of type II, and 3 of type III, we might calculate the probability of drawing a sample of 6 that contained exactly 3 type I, 2 type II, and 1 of type III. The hypergeometric term would be

$$P(3\text{I}, 2\text{II}, 1\text{III}, 12, 5, 3) = \frac{\binom{12}{3}\binom{5}{2}\binom{3}{1}}{\binom{20}{6}} = \frac{55}{323}$$

Note that the numerator is the number of ways that 3 type I, 2 type II, and 1 type III can be drawn from their populations of 12, 5, and 3 respectively. This number is 6600. The denominator is merely the number of ways the draw of 6 can be made from 20.

Given a population containing a finite number of items w, divided into k exhaustive and mutually exclusive classes each containing w_k, then the probability of occurrence of a sample containing a specified number r_i in each class w_i is

$$P(r_1, r_2, \cdots r_k)/w_1, w_2, \cdots w_k) = \frac{\binom{w_1}{r_1}\binom{w_2}{r_2} \cdots \binom{w_k}{r_j}}{\binom{w}{r}}$$

where $w_1 + w_2 + \cdots w_k = w$ and $r_1 + r_2 + \cdots r_k = r$.

As the population size w approaches a large value, the probabilities found with the hypergeometric will approach the multinomial or binomial. The mean of the hypergeometric with only two classes is equal to the binomial, np.

13

POISSON DISTRIBUTION

In Chapter 10, we examined the binomial distribution as it described a Bernoulli process. We now wish to consider another discrete distribution that will provide probabilities for situations similar to that for which the binomial is used. However, there are some differences from the Bernoulli experiment.

Instead of a finite number of trials resulting in a finite number of terms in the distribution, the process will operate over continuously over a large number of trials.

Rather than being able to count both successes and failures in the outcomes of trials, the process produces only successes or occurrences. These occur randomly throughout the trials but are governed by a mean value of occurrence or an occurrence rate.

For example, we may visualize a production line producing television sets. We want to know the probability of some number of failures in these sets. Such failures may occur in a single set or in many sets. A series of terms describing the number of failures that could occur in a large number of television sets operating over many exposure hours would be nearly infinite. Likewise, the number of accidents that could occur in a large number of aircraft flight hours would approach a large number and would be a good candidate for solution by this type of distribution.

The distribution was named after Simeon Poisson (1781–1840), a French mathematician. It may be applied only to a process that may be thought of as dichotomous, stationary, and independent as with the binomial. Poisson derived the *probability mass function* for this discrete distribution as a limiting form of the binomial with n approaching infinity while holding np constant.

The Poisson process is a particular form of the Markov process. In a Poisson process:

The expected number of occurrences per unit exposure remains constant throughout any period of exposure.

The likelihood of occurrences remains independent.

Only occurrences of the specified event or lack of them will be considered.

The probability mass function as derived by Poisson is the following:

$$f(x) = \frac{(\lambda t)^r \, e^{-\lambda t}}{r!} \qquad \text{if } r = 0, 1, 2, \cdots \qquad (13.1)$$

where e = constant, approximately 2.7183, λ = mean occurrence rate per unit exposure, and t = units of contiguous exposure.

The cumulative probability mass function for the Poisson distribution may be derived as follows. Assume that we wish to develop an infinite series for which the sum of all terms is equal to 1.00 and each term equals the probability of exactly some number of occurrences, 0, 1, 2, $\cdots$

Expanding e^a in an infinite series:

$$e^a = 1 + \frac{a}{1!} + \frac{a^2}{2!} + \frac{a^3}{3!} + \cdots \qquad (13.2)$$

Multiplying this series by e^{-a}, we have:

$$1 = e^{-a} + \frac{a e^{-a}}{1!} + \frac{a^2 e^{-a}}{2!} + \frac{a^3 e^{-a}}{3!} + \cdots \qquad (13.3)$$

Substituting $\lambda t = a$ and rearranging terms:

$$e^{-\lambda t} + \frac{\lambda t e^{-\lambda t}}{1!} + \frac{(\lambda t)^2 e^{-\lambda t}}{2!} + \frac{(\lambda t)^3 e^{-\lambda t}}{3!} + \cdots = 1.00 \qquad (13.4)$$

or

$$P(0) + P(1) + P(2) + P(3) + \cdots = 1.00$$

where λ = rate of event occurrence per unit exposure and t = exposure period in units homogeneous with λ.

Note that λ is an event rate as opposed to p of the binomial, which is a probability of a single occurrence per unit exposure. Thus, λ may take values greater than 1.00.

Further note that each term of equation (13.4) is equal to equation (13.1) with $r = 0, 1, 2, \cdots$. Because of the derivation of equation (13.1), each term is only approximately equal to the binomial probabilities. However, the approximation will be quite good if $t/\lambda \geqslant 200$.

If the Poisson distribution is used for a system in which the number of event occurrences is limited, another approximation is introduced because of the infinite nature of the series. A moments investigation with a calculator will disclose that the probability space represented by a very large number of event occurrences is sufficiently small to maintain acceptable accuracies in computations using the Poisson distribution for situations where a finite number of events may occur.

The Poisson distribution has attributes that make it most useful in safety. It should be considered in greater detail than the other distributions.

Let us assume that an electric component has been tested with a replacement test and it is found that 20 failures occur in 15,000 hours of testing. If the testing is done by the reliability department, the results will most likely be reported as a mean time to failure (MTBF). We designate this value as m. Noting the dimensions of m as hr/f, we can readily see that it is the reciprocal of λ, which has units f/hr. Thus, a Poisson expression for the probability of exactly two failures in 800 hours could be written as:

$$P(2f)_{800} = \frac{[(1.333E-3)(800)]^2 e^{-(1.333E-3)(800)}}{2!} = .1958$$

where $\lambda = 1.333E-3 = 20/15{,}000\ f/hr$.

Likewise, the probability of two failures may be written as

$$P(2f)_{800} = \frac{(800/750)^2 e^{-(800/750)}}{2!} = .1958$$

where $800/750 = t/m$

13.1 EXPONENTIAL DISTRIBUTION

The exponential and Poisson distributions are quite similar. Both address a Poisson process in which λ, the event rate is constant and stationary. However, the exponential distribution is continuous whereas the Poisson is discrete. The exponential distribution is the distribution of time to event occurrence. It will also yield the probability of at least one failure in an exposure time, t. Thus, the exponential distribution is perhaps more useful that the Poisson in safety calculations. In continuous distributions the point value is called the probability density function. The probability density function of the exponential distribution is

$$f(x) = \lambda e^{-\lambda t} \tag{13.5}$$

From equation (13.1) we can see that the probability of zero event occurrences is computed when $r = 0$.

$$P(0)_t = e^{-\lambda t} \tag{13.6}$$

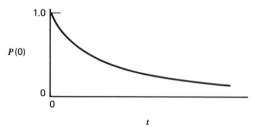

Figure 13.1 Probability of zero event occurrence.

Thus, the probability of at least one event in t is

$$P(\geq 1f) = 1 - e^{-\lambda t} \qquad (13.7)$$

In the above example, the probability of at least one failure in 800 hours is

$$P(\geq 1f) = 1 - e^{-1.067} = .6558$$

where $1.067 = $ expected number of events in t hours.

The nature of the exponential distribution may be further revealed by examining the term of the distribution, $e^{-\lambda t}$ or the probability of exactly zero event occurrences in t. Figure 13.1 illustrates.

Figure 13.1 provides a simple model of the failure of electromechanical systems. We can see that the probability of zero failures is a certainty or 1.00 if we do not expose the system with t. This assumes no shelf or storage failures. Then, as we expose the system to its operational environment, the inherent failure rate of the component manifests itself. The probability of zero failures will continually decrease with exposure, approaching asymptotically zero, at a large component exposure. In Figure 13.1, the value of λ along the curve is a constant.

In such a simple and attractive relationship, we might be tempted to carry the mathematics beyond its limit. The limit of extrapolation of the curve is that amount of exposure, t, that was placed on a single component in establishing the λ. For instance, in the replacement test of the foregoing example, if 10 items were each tested 1500 hours for a total of 15,000 hours, then the λ so determined as $1.333E-3$ may not be used beyond 1500 hours without additional testing, certainly not to 15,000 hours.

For each different λ that is determined by testing, there will be another curve. Figure 13.2 illustrates a family of such curves.

If we select one of the components represented by one of the curves of Figure 13.2 and plot its complement as in Figure 13.3, we can illustrate another facet of the exponential distribution.

Note that the complementary curve to the $P(0)$ is the $P(\geq 1)$. This curve illustrates a disturbing property of the Poisson and, in fact, the failure process that it describes. This is that given enough exposure it is nearly a certainty that at least one failure

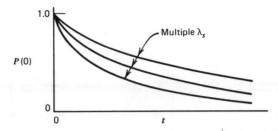

Figure 13.2 Probability of zero event occurrences (multiple λs).

will occur no matter how small the failure rate, λ, has become. This means that if you put enough components in service to accumulate exposure, or keep one component in service for a extreme amount of time, the failure will almost certainly occur. This may be thought of as a proof of Murphy's law that in its general form says, "whatever can go wrong, will." This is even more disturbing when we consider not a small component, but the failure of a major system at a prohibitive loss.

From Figure 13.3, we can see that to prevent an accident with a certainty, redesign is the only solution. Procedures, warning devices, and safety devices will only result in a reduced λ and an extension of the amount of exposure that the system will sustain until the undesired event nears certainty. Thus, we see that Murphy's law is more than a rueful acknowledgment of the worst seeming always to happen; it is the inexorable functioning of the laws of probability.

Figures 13.1 to 3 above assume constant λ. This would imply that the component does not wear out over the exposure for which computations are to be made. It will be found that if a large number of components are tested to determine λ, the failure rate over time will be as shown in Figure 13.4.

In Figure 13.4, we can see the three zones of component life. Zone I is the "wear-in" or "burn-in" phase. A few of the components that are being tested are not at the quality standard designed into the component due to manufacturing defects. These components will fail early in the testing program causing a higher than expected mean λ early in testing. The failure and removal of these components

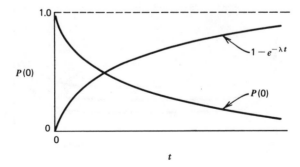

Figure 13.3 Probability of one or more events occurring.

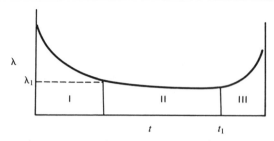

Figure 13.4 Failure rate bathtub curve.

from the replacement test will facilitate finding the λ_1 for zone II or the operational region of component life.

As testing continues, the component will be begin to wear out and λ will increase. The limit of zone II, t_1, is of considerable importance. It is this amount of exposure, beyond which we can not extrapolate the Poisson computations because of the increasing λ. Zone III is the wear-out region. The component should be replaced or refurbished before operating in this region.

For instance, if we have an electrical generating system with a voltage regulator operating continuously, we may wish to calculate the probability of voltage regulator failure in a year of operations given an MTBF(m) of 12,000 hours. This m has been determined from a replacement test and would correspond to λ_1 in Figure 13.4. We could calculate the desired probability as follows:

$$P(\geqslant) = 1 - e^{-8700/12,000} = .5157$$

where 8700 = number of exposure hours in a year.

Although a probability was easily calculated, there remains the question of 8700 hours being less than t_1. If 8700 hours was greater than t_1, then such a calculation would have been improper because the m at 8700 hours would have been less than 12,000 hours.

As a matter of interest, note that if t_1 was, indeed, more than 8700, the expected number of failures of the voltage regulator in a year is .725. This number may be

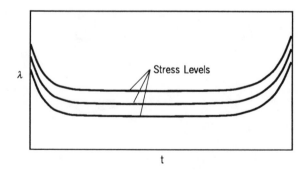

Figure 13.5 Failure rates under varying stress.

of greater assistance in determining the state of safety than the probability of at least one failure in a year.

The bathtub curve of Figure 13.4, may be stress sensitive. If the curve describes the λ of an electrical component, overstressing the component may cause an increase in all points of the curve. A different curve would exist at different stress levels. In this case, a family of curves would more properly describe the λ of a component. Figure 13.5 illustrates. Types of stress that may cause variations in the bathtub curve of λ may be, electrical load, vibration, k temperature, or static or dynamic loads.

The exponential distribution is also useful when it is necessary to determine the event rate or the allowable exposure to meet a probabalistic standard. Suppose that in the example of an electrical generating system we wish a probability of at least one failure in a year of operations to be no greater than .3. There would then be the question of the needed m or λ of the voltage regulator to meet the desired standard. A simple statement of the problem would be equation (13.8).

$$P(\geqslant) = 1 - e^{-\lambda(8700)} = .3 \qquad (13.8)$$

We may transform this equation to a logical equivalent statement.

$$P(0) = e^{-\lambda(8700)} = .7$$

Taking the natural log of both sides of the equation and bringing down the exponent of the left side:

$$-\lambda(8700) \ln e = \ln .7$$

Noting that $\ln e = 1.00$ and solving for λ:

$$\lambda = \frac{-\ln .7}{8700} = 4.100E-5$$

We can see that this is less than the given λ:

$$\lambda = 1/12{,}000 = 8.333E-5$$

If it is desired to find the number of exposure hours, t, at which the desired probability of .3 would be exceeded given the original m, a similar computation could be performed.

Unfortunately such a calculation does not differentiate between systems. The exposure value is total exposure on all systems. Thus, if there are two electrical generating systems to be considered, the exposure time of 8700 hours would be exceeded in one-half a year and these calculations would apply to only one-half year of calendar time.

14

NORMAL DISTRIBUTION

In previous sections we have examined several discrete and one continuous distributions. In continuous distributions, as was noted in the exponential, the value X may take on any value or an infinite number of values. Thus, the concept of computing the probability of X taking on a value x is not useful. The probability of X taking on any specific value x in a continuous distribution is 0.0 or $P(X = x) = 0.0$. Thus, with a continuous distribution, it is meaningful only to calculate probabilities over an interval of X. If one is concerned with a variable that must be represented by a continuous distribution, a probability may be only calculated between limits of values of the variable.

$$P(a \leqslant X \leqslant b) = \int_a^b f(x)\, dx \qquad (14.1)$$

The function $f(x)$ can be viewed as the mathematical function that describes a curve of a continuous variable. Thus, the value of $f(x)$ evaluated at any point X is merely the ordinate of the curve, not a probability. The properties of $f(x)$ in a continuous distribution are the following:

1. $f(x) \geqslant 0$ for all real values of X; this means that $f(x)$ cannot be negative for any value of X between $-\infty$ and $+\infty$.
2. $\int f(x) = 1.0$ between $-\infty$ and $+\infty$; that is, the total area under the curve described by $f(x)$ must be 1.0.

Corresponding to the cumulative distribution function of a discrete variable, we have the cumulative distribution function of a continuous distribution. As in the case of a continuous distribution, this function represents the probability that the variable X is no greater than some value x. It must be found by integration of the probability density function.

$$F(x) = P(X \leqslant x) = \int_{-\infty}^{x} f(X')dx' \qquad (14.2)$$

where $f(x')\, dx'$ is used in place of $f(x)\, dx$ as a dummy variable because x is the upper limit of integration.

The normal distribution is a continuous distribution although it may be used to approximate variables that are discrete. The distribution is sometimes called the Gaussian distribution after Carl Gauss, although the distribution was truly described by Alexander Demoivre in 1733. The distribution is fixed by two parameters σ, the standard deviation, and μ, the arithmetic mean. Figure 14.1 illustrates the distribution.

The probability density function is given by

$$f(x) = \frac{1}{2\pi\sigma^2}\, e^{-(x - \mu)^2/2\sigma^2} \qquad (14.3)$$

where

$$\mu = \text{arithmetic mean of the distribution}$$

$$\sigma = \text{standard deviation}$$

The density function of equation (14.3) describes a continuous bell-shaped curve as shown in Figure 14.1. The curve is symmetrical about μ at the point where the function $f(x)$ has the maximum ordinate. Thus, the arithmetic mean, median, and mode have the same value. Because of this symmetry, $P(X \leqslant \mu) = P(X \geqslant \mu) = .50$. The curve mathematically extends from $+\infty$ to $-\infty$ and is asymptotic to the horizontal axis. We can see from equation (14.3) that it is completely determined by the parameters σ and μ.

This distribution is not of particular interest to us for values of the probability density function $f(x)$. Our interest in this useful and common distribution lies in using it to calculate areas under an interval of the curve that will represent probabilities of the variables taking on any value over this interval, or in some cases calculating the cumulative density function. These two calculations are expressed as follows.

$$P(a \leqslant X \geqslant b) = \int_{a}^{b} f(x)dx = \int_{a}^{b} \frac{1}{2\pi\sigma^2}\, e^{-(x-\mu)^2/2\sigma^2} \qquad (14.4)$$

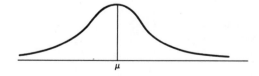

Figure 14.1 Normal distribution.

$$P(X \leq a) = \int_{-\infty}^{a} f(x)dx \qquad (14.5)$$

The direct integration of equations (14.4) or (14.5) would be cumbersome. It is possible to avoid this direct integration for manual calculations by providing a table in which a standardized normal function has been integrated. The normal density function may be placed in standardized form as follows:

$$f(z) = \frac{1}{2\pi} e^{-z^2/2} \qquad (14.6)$$

where $z = \dfrac{x - \mu}{\sigma}$

If equation (14.6) is integrated for various values of z, these values may be stored in a table for evaluation of integrals of a normal density function given a μ and σ. Table A.1 is such a table for one-half of the curve. Since the curve is symmetrical, knowing areas under one-half allow one to find any area in question.

It is common to talk of parameters that fall within 1, 2, or 3 standard deviations on either side of the μ as the 1, 2, or 3σ limits of values of an event though the interval in which they lie is 2, 4, or 6σ wide.

The following example will illustrate the use of this distribution. Let us assume that we are concerned about a pressure relief switch that we are to place in our system to protect it from an overpressure condition. Testing switches from a production line, it is found that the probability density function of the pressure at which the switches open is normal. This can be done through a test such as the chi-square. The pressure at which the switches activate has a μ of 60 pounds per square inch (psi) and a σ of 12 psi. Figure 14.2 illustrates the distribution of the pressure at which this switch activates.

Note in the above figure that if the μ of the distribution is changed, the entire curve will slide along the x axis of the real world dimension X, in this case, psi. If the sigma of the curve is increased the curve will flatten in place. If the sigma is reduced, the curve will stand up more narrowly, although always retaining an enclosed area of 1.0.

The picture of a normal curve is useful in determining how the calculation is to be made. To determine areas under the positive portion of the curve, it is necessary

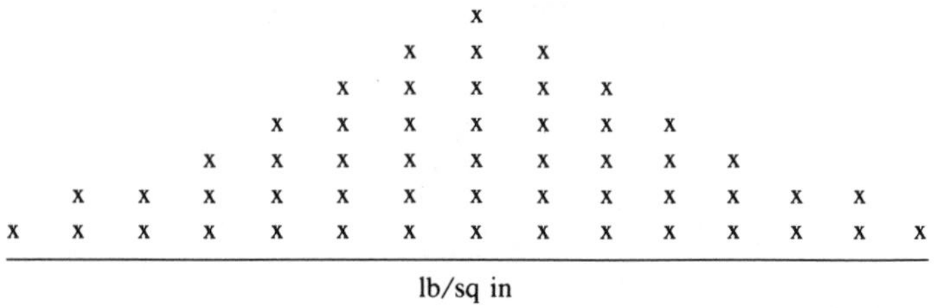

```
                        x
                  x     x     x
            x     x     x     x     x
      x     x     x     x     x     x     x
      x     x     x     x     x     x     x     x
x     x     x     x     x     x     x     x     x     x     x
x     x     x     x     x     x     x     x     x     x     x     x     x
```

lb/sq in

Figure 14.2   Pressure switch histogram.

to enter the table at z values or the number of standard deviations a point is from the mean. A z value may be calculated from the following relation.

$$z = \frac{X - \mu}{\sigma} \qquad (14.7)$$

In the example of the pressure relief switch, let us assume that the maximum pressure that the system will take to rupture is 72 psi., Thus, we would be concerned with the probability that a pressure relief switch with a normal distribution of its opening pressure with a μ of 60 psi and σ of 12 psi will activate at a pressure greater than 72 psi. This probability will be

$$P(X > 72) = 1 - \int_{72}^{\infty} f(x)dx$$

where $f(x)$ = equation (14.3).

Calculation of the z value with which to enter Table A.1:

$$z = \frac{72 - 60}{12} = 1.00$$

From Table A.1:

$$P(72 \geqslant X \geqslant 60) = .34135$$

Therefore, the probability that $X > 72 = 1 - (.5 + .34135) = .15865$.

There is, of course, another question concerning this pressure switch. Assuming that the normal operating pressure of the system into which the switch will be inserted is 41 psi, there remains the question of the switch opening at normal operating pressure and preventing the system being brought on line. This probability may be calculated as follows:

$$P(X < 42) = 1 - P(42 \leqslant X \leqslant 60)$$

$$z = \frac{42 - 60}{12} = -1.5$$

$$P(42 \leqslant X \leqslant 60) = .43320$$

Therefore, the probability that $X < 42 = 1 - .43320 = .56680$.

A final question might arise concerning this pressure switch. What is the probability that any switch from this production line will be satisfactory?

$$P(\text{Sat SW}) = P(42 < \text{SW} \leqslant 72) = .43320 + .34135 = .77455$$

This value is not large and would undoubtedly cause the safety analyst to judge the switch unacceptable. In many cases, the calculations made with the normal density function are approximate. It is clear, for instance, that a pressure switch would not function at a pressure less than zero, although the area of the lower tail of the distribution down to $-\infty$ is included in the integration. However, this error is very slight and can be ignored in calculations of this nature. The same reservations may be made for calculations that include the area of the upper tail to $+\infty$.

15

LOGNORMAL DISTRIBUTION

The logarithmic normal distribution is useful in describing safety parameters that may vary due to nonuniform factors. Such a distribution will be skewed positive. As the name implies, the lognormal distribution has a simple logarithmic relation to the normal distribution. The transformation occurs through taking logarithms of the variable x, and finding the mean and standard deviation of the logarithms.

Thus, the probability density function is

$$f(x) = \frac{1}{x\sigma(2\pi)^{1/2}} e^{-1/2(f(u))^2} \tag{15.1}$$

The cumulative distribution function is

$$F(x) = \frac{1}{\sigma(2\pi)^{1/2}} \int_0^x f(x)dx \tag{15.2}$$

where

$$f(u) = \frac{\ln x - \mu_1}{\sigma_1}$$

x = value of random variable

μ_1 = mean of the logarithms of variable x

σ_1 = standard deviation of logarithms of x

145

The lognormal cumulative distribution function is an S curve of a similar shape to the normal cumulative distribution function. Quite simply, the lognormal is the normal distribution with $f(u)$ substituted for z values.

The lognormal distribution may be used to describe event rates subject to wearout (λ) and repair or mean down times (τ). It is also useful in describing human error rates. Because the calculation of values of the distribution is cumbersome due to the necessity of taking logarithms of many values, the practical use of the distributed is facilitated by a Monte Carlo program that will calculate the variables quickly. Such a program is available for a microcomputer. The lognormal distribution takes the following shape (Figure 15.1).

The mean and variance of the lognormal distribution are the following.

$$E(x) = e^{\mu_1 + (1/2)\sigma_1^2} \tag{15.3}$$

$$V(x) = e^{2\mu_1 + \sigma_1^2}(e^{\sigma_1^2} - 1) \tag{15.4}$$

Assume that a value τ is distributed as the lognormal distribution with a mean of 5 hours. Thus:

$$z = \ln 5 = 1.6094$$

$$z = 1.6094 = \mu_1 + (1/2)\sigma_1^2$$

$$\text{Assume that } \sigma^2 = .2188$$

$$\mu_1 = 1.5$$

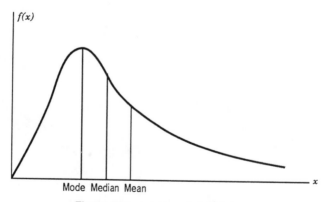

Mode Median Mean

Figure 15.1   Lognormal distribution.

To determine $P(x \leq 8)$, we use the following transform.

$$P(x \leq 8) = P\left(Z \leq \frac{\ln 8 - 1.5)}{(.2188)^{1/2}}\right) = P(Z \leq 1.23)$$

From the normal table:

$$P(Z \leq 1.23) = .8907$$

16

WEIBULL DISTRIBUTION

The Weibull is named after W. Weibull of Sweden, who first published its application to fly ash and strength of materials in 1951. It was popularized by J.H.K. Kao in 1957 when he applied it to the failure of electronic components and systems. The distribution is a three parameter distribution that can be made to approximate many of the continuous distributions that are so useful.

The probability density form of the Weibull distribution is:

$$f(x) = \frac{\beta}{\alpha} \left(\frac{x - \gamma)}{\alpha} \right)^{\beta-1} \exp \left[- \left(\frac{x - \gamma)}{\alpha} \right)^{\beta} \right] \qquad (16.1)$$

where α is the scale parameter; β is the shape parameter, > 0; γ is the location parameter; and α, $\gamma \leq x$.

The cumulative distribution function is

$$F(x) = 1 - \exp \left[- \left(\frac{x - \gamma)}{\alpha} \right)^{\beta} \right] \qquad (16.2)$$

If $\beta = 1$, equation (16.1) becomes the exponential distribution probability density. The distribution can be made to approach the normal if β increases towards 4 with γ and α fixed.

If one is to use the Weibull distribution to describe event occurrence data, it is necessary to estimate the three parameters. If the location parameter γ is set equal to zero, it would allow the events to begin to occur at $t = 0$. This should be common practice for fault events as it would establish a threshold of exposure before which no events may occur and assume events could occur immediately after exposure is initiated.

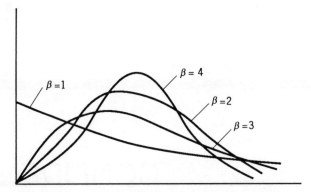

Figure 16.1 Variation of Weibull shapes ($\gamma = 0$, α fixed).

The shape, location, and scale parameters do not have direct counterparts to the mean and variance of two-parameter distributions previously examined. Figure 16.1 below illustrates the effect of the shape parameter on the distribution.

Weibull variates are best calculated using a computer. It would be most helpful to have a Monte Carlo program to assist in the calculation and application of the distribution to a system analysis problem. It can be used to approximate event–occurrence rates from human to component.

A simple example of manual calculation of Weibull values is an electronic subsystem known to have a Weibull density with $\gamma = 0$, $\beta = 1/2$, and $\alpha = 100$. The fraction of the subsystems expected not to survive to 600 hours is:

$$1 - F(600) = 1 - e^{-(600/100)^{1/2}} = .9137$$

17

CONFIDENCE
LIMITS

If we conduct two replacement reliability tests of a component in an effort to determine its failure rate (λ), we would be very surprised to find the same failure rate on each test. Our awareness of the manner in which components fail will result in an expectation of variation in this rate as in all other rates of event occurrence that are governed by chance causes. Thus, repeated samples of the accident rate of a system would result in a rate that would vary from sample to sample.

The manner in which failure or accident rates develop indicate that multiple samples of a rate are needed if one is to estimate a value of a rate to be used in safety analysis. Multiple samples would allow an estimate of the mean and variance of the rate and thus some statistical method could be used to find a *conservative* estimate of a rate to be used in analyzing a system. To be more specific, multiple samples would allow the analyst to estimate an interval in which the true mean of the rate would lie with some probability of being correct in this estimate.

It can be shown that the mean of a sample is the best estimator of the true mean. If repeated samples of a population are taken, then the mean of the sample means will continue to be the best estimator of the true population mean. However, in either case, the probability that the true mean will be exactly the sample mean, will approach zero and the probability that the true mean will be greater than or less than the sample mean will be approximately .5 if the distribution of mean is not highly skewed. If one were seeking a conservative estimate of an event rate, a mean would not be the proper one to choose. A conservative estimate would properly be an upper bound on the region in which we believe the true mean rate to lie. Statistical methods will allow us to place a probability on the likelihood of the true mean rate being within this region.

17.1 NORMAL DISTRIBUTION CONFIDENCE INTERVALS (KNOWN σ)

If the underlying population distribution of a variable is known to be normally distributed and the standard deviation (σ), of this distribution is known, it is possible to predict with a probability the interval in which the true population mean will lie. Such a prediction may be developed by equation (17.1).

$$P(z_1\sigma_{\bar{x}} \leq \mu \leq z_2\sigma_{\bar{x}}) = 1 - \alpha \qquad (17.1)$$

where

z_1 = the number of σs from the mean to the lower limit of the interval
z_2 = the number of σs from the mean to the upper limit of the interval
z = $\dfrac{x - \bar{x}}{\sigma}$

$\sigma_{\bar{x}}$ = the magnitude of the standard deviation of the distribution of sample means estimated by $\sigma/\sqrt{n}$
α = risk in making such a prediction where
σ = the population standard deviation
n = sample size

Figure 17.1 illustrates an interval on the normal distribution.

Admittedly, it is unlikely that one would know the true σ of a population and not know the true mean. The foregoing discussion is offered as an introduction to the more conventional case of interval estimation.

17.2 *t* DISTRIBUTION CONFIDENCE INTERVALS (UNKNOWN σ)

If we sample repeatedly from a normally distributed population for which the true mean and standard deviation are not known, we must use a distribution first developed by W.S. Gossett in 1908 under the name "Student." It has come to be known as the Student's *t* distribution or simply the *t* distribution. The distribution has a density

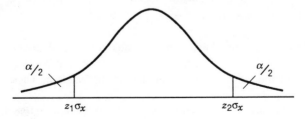

Figure 17.1 Confidence interval estimation with normal distribution.

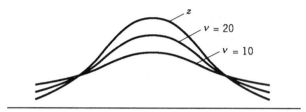

Figure 17.2 Comparison of the standard normal with t distribution for different degrees of freedom.

function quite different than the normal distribution. However, the shape of the distribution can be similar to the normal. It is symmetrical about a mean of zero, however, the variance of distribution is not constant but varies with a single parameter—degrees of freedom (v). As the degrees of freedom increase, the t distribution approaches the normal. It is generally considered sufficiently identical at a v of 30 to be considered normal. At smaller values of v, the distribution has a larger variance and develops thicker tails than the normal distribution.

Figure 17.2, illustrates the differences in the t from the normal.

The probability density function of the t distribution is:

$$f(x) = g(v)\left(1 + \frac{t^2}{v}\right)^{-(v+1)/2} \tag{17.2}$$

where $g(v)$ is a function of v and a constant.

If σ of the underlying population is not known, the t distribution can be used to estimate confidence intervals. To do so, it is necessary to calculate a t value that will correspond to a z value when confidence intervals are calculated from the normal distribution.

$$P(t_1 s_{\bar{x}} < \mu < t_2 s_{\bar{x}}) = 1 - \alpha \tag{17.3}$$

where

t_1 = the number of ts from the mean to the lower limit of the interval,
t_2 = the number of ts from the mean to the upper limit of the interval
$s_{\bar{x}}$ = is an estimate of the standard deviation of the distribution, $s_x = s/n$ where
s = the standard deviation of the sample

The t values may be found in a standard t table for the probability or risk required in a tail. They may also be calculated with a programmable hand calculator. The shape of a t distribution depends on the degrees of freedom. These are $n - 1$ for the given sample.

In safety, the analyst is primarily concerned with a conservative or upper bound on the rate. Thus, if all the risk is placed in the upper tail of the t, the following relation would obtain.

$$P(\lambda \leq t_\alpha \sigma_{\bar{x}}) = 1 - \alpha \qquad (17.4)$$

This extremely brief treatment of confidence intervals for the normal and t distribution is intended only as background for a method that is most useful to safety practitioners.

17.3   CHI-SQUARE CONFIDENCE INTERVALS

We have examined methods of estimating the interval in which the true mean value of a parameter may lie that require multiple samples. In safety, it is unusual to take multiple samples to determine a state of safety or to predict future safety performance. The more common approach to measuring safety performance is to take a single sample based on exposure, determine an accident rate, and use this rate to make extrapolations for future safety performance. The rate (λ) so determined is the best estimate of the true mean rate, however, it is not a satisfactory value with which to predict future performance. As has been observed, there is an approximate probability of .5, that the true mean rate will be greater than this single sample value.

It was shown by Epstein (1960) that the mean time between failure [MTBF, m] can be approximated by the chi-square distribution if m is exponentially distributed. This opened the way for a very simple method of finding conservative confidence bounds on accident or event failure rates that have an exponential distribution.

The sum of the squares of standardized normal deviates may be designated as χ^2.

$$\chi^2_{(n)} = \sum_{}^{n} \frac{(x_i - \mu)^2}{\sigma^2} = \Sigma z_i^2 \qquad (17.5)$$

The distribution of this random variable has a form that depends on the number of independent observations. In the case of χ^2, this number is the number of degrees of freedom v. The effect of v on the shape of the distribution is shown in Figure 17.3.

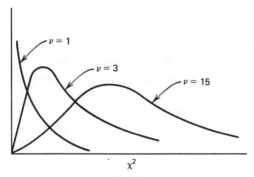

Figure 17.3   Chi-square distribution.

The probability density function of this distribution is

$$f(\chi^2) = f(\nu) \exp\left(-\frac{\chi^2}{2}\right)(\chi^2)^{(\nu/2)-1} \quad \text{for } \chi^2 \geqslant 0, \ \nu > 0 \quad (17.6)$$

where $f(\nu)$ is a function of degrees of freedom.

The mean and variance of the χ^2 distribution are

$$E(\chi^2_{(n)}) = n \tag{17.7}$$

$$V(\chi^2_{(\nu)}) = 2\nu \tag{17.8}$$

Epstein was able to show that

$$2F(\hat{m}/m) \tag{17.9}$$

has a chi-square distribution with $2F$ degrees of freedom.
where

F = number of events that have occurred in a given exposure, T
$\hat{m}$ = observed mean time between events
m = true mean time between events

Therefore, we can make a probability statement from equation (17.9).

$$P\left(\chi^2_{1-\alpha/2;2F} < \frac{2F\hat{m}}{m} < \chi^2_{\alpha/2;2F}\right) = 1 - \alpha \tag{17.10}$$

This expression states that the true value of m will lie within the range of χ^2 values with confidence $1 - \alpha$. Rearranging terms of equation (17.10), the following expression is in more workable form.

$$P\left(\frac{2F\hat{m}}{\chi^2_{\alpha/2;2F}} < m < \frac{2F\hat{m}}{\chi^2_{1-\alpha/2;2F}}\right) = 1 - \alpha \tag{17.11}$$

Epstein was also able to show that when the test was conducted to determine m that was not stopped on a specified number of event occurrences or failures, the degrees of freedom for the lower bound on m in equation (17.11), would be $2F + 2$. In addition, noting that $2F\hat{m} = T$, where T = total exposure in the sample, equation (17.11) may be further rewritten as

$$P\left(\frac{2T}{\chi^2_{\alpha/2;2F+2}} < m < \frac{2T}{\chi^2_{1-\alpha/2;2F}}\right) = 1 - \alpha \tag{17.12}$$

Finally, observing that

$$\lambda = 1/m$$

the statement of the bounds on the true mean value of λ may be put in its final form for the two-tailed bound.

$$P\left(\frac{\chi^2_{1-\alpha/2;2F}}{2T} < m < \frac{\chi^2_{\alpha/2;2F+2}}{2T}\right) = 1 - \alpha \qquad (17.13)$$

However, in safety, unlike reliability, we are normally concerned only with the conservative bound on a parameter. In the case of λ, this would be the upper bound. Thus, the expression for the conservative bound on a rate would be the following equation:

$$P\left(\lambda < \frac{\chi^2_{\alpha;2F+2}}{2T}\right) = 1 - \alpha \qquad (17.14)$$

An example will illustrate the use of equations (17.13) and (17.14). Let us assume that aircraft accident data from which rates may be derived will approximate the exponential distribution as it usually is assumed to do. Assume that the data show that five accidents occurred in 300,000 hours of flying. The best estimate of the true mean rate from this data is, of course, $1.667E-5$ per hour. Let us see how this compares to the 90% confidence bounds on the rate.

From equation (17.13) we may write the following statement:

$$P\left(\frac{\chi^2_{.95;10}}{2(3E5)} < \lambda < \frac{\chi^2_{.05;12}}{2(3E5)}\right) = .90$$

Note that the statement excludes two tails from the confidence interval. These tails contain .1 of the total area under the distribution.

Solving the interval bounds we find:

$$P(3.940/6E5 < \lambda < 21.026/6E5) = .9$$

$$P(6.567E-6 < \lambda < 3.504E-5) = .9$$

Thus, we can see that given the data we would expect the true mean rate to be between $6.567E-6$ and $3.504E-5$ with a .9 probability. To state another way, there is a risk of .1 that these bounds would not include the mean rate.

Without further calculation, we may make another statement:

$$P(\lambda < 3.504E - 5) = .95$$

This statement simply includes the lower tail in the confidence area. It is the latter statement that would normally be made in safety situations. The latter statement may be arrived at directly from equation (17.14).

Using the Poisson term, we may then calculate the probability of no accidents in future exposure t with 95% confidence. Letting $t = 50,000$ hours:

$$P(0)_{50k} = e^{-3.504E-5(50k)} = .1734$$

As .173 is a *conservative* confidence bound, we may say that there is a .173 probability of there being no accidents in 50,000 hours and there is 95% confidence that the true value will be less than that value. Of course, being less than is better than.

We may also calculate the probability of any or at least one accident in 50,000 hours with the same confidence.

$$P(\geq 1)_{50k} = 1 - .1703 = .8266$$

We have seen that the chi-square distribution becomes less skewed as the degrees of freedom (ν) increase. At or above 30 degrees of freedom, it is considered that the chi-square distribution is essential the same as the normal distribution. For this reason, tables of the chi-square distribution seldom go above 30 degrees of freedom. Fischer (1956) showed that when $\nu \geq 30$, the expression $2\chi^2$ not only forms a normal distribution but has a mean of $2\nu - 1$. Thus, the following expression may be formed:

$$\sqrt{2\chi^2} = \sqrt{2\nu - 1} + z \tag{17.15}$$

Solving equation (17.15) for χ^2 we have:

$$\chi^2 = 1/2\,(z + \sqrt{2\nu - 1})^2 \tag{17.16}$$

Equation (17.16) allows us to solve for χ^2 by selecting z from the normal table, depending on the size of the tail one wished to cut for the specified α. However, since beyond 30 degrees of freedom the z values will remain constant for each α specified, appropriate z_α may be added to the end of each column of the χ^2 column. This has been done in Table A.2 at the end of this book.

EXERCISES FOR PART II

Probability

1. A box contains 10 balls. Five are white, 3 are black, and 2 are red.
   (a) What is the probability that the first and third balls drawn from this box without replacement will be black?
   (b) What is the probability that exactly 2 black balls will be drawn in three draws without replacement?
   (c) What is the probability that at least 2 black balls will be drawn in three draws without replacement?
   (d) What is the probability that 2 white and 1 red ball will be drawn in four draws without replacement?

2. Do problems 1 to 4 with replacement.

3. One must select from a set of 12 components that are known to contain 4 faulty components. The selection is made at random so that a selection of any component is equally likely.
   (a) What is the probability of drawing exactly 1 faulty component if 3 different components are drawn?
   (b) What is the probability of drawing exactly 2 faulty components if 3 different components are drawn?
   (c) What is the probability of drawing at least 2 faulty components if 3 different components are drawn?
   (d) What is the probability of drawing any faulty component if 3 different components are drawn?
   (e) What is the probability of drawing at least 1 faulty component in either of two draws if 3 different components are drawn in each of the two draws (i.e., no components are replaced)?
   (f) If 5 components are drawn from two identical sets of 12 components each assumed to contain 4 faulty components, what is the probability of at least 1 faulty component in the two sets of draws?
   (g) If 5 components are drawn from 20 identical sets of components, what is the probability of at least 1 faulty component in any of the draws?

4. There is concern for the vulnerability of a ship. It is divided into eight regions in which it may be hit by a missile, each having a vulnerability measure. There are two fatal regions of the eight. There are three regions that will be seriously damaged if hit. Hits in the remainder of the regions will cause minor damage. For simplicity we will assume that the probabilities of a missile hitting any region are equal.
   (a) What is the probability of a fatal hit, given that the ship is hit?
   (b) What is the probability of a fatal hit if the ship is hit twice?
   (c) If a ship is hit once, what is the probability of serious or fatal damage?

(d) In the course of an engagement of five of these ships, three are hit once. What is the probability of at least one fatal hit?

(e) If one of the fatal regions is armored so that it becomes a serious damage region, what is the probability of at least one fatal hit under the conditions of 4(c)?

5. A large aerospace system contains five major subsystems. Failure of either of two of the subsystems will result in total system failure. Failure of any one of the other three will reduce system effectiveness. Failure of two of the other three will result in total system failure. Since no information is available as to the likelihood of failure of these subsystems, it is assumed that their failure is equally likely.

(a) If one subsystem fails, what is the probability of complete system failure?

(b) If two subsystems fail, what is the probability of complete system failure?

(c) If two subsystems fail, compare the probability of system failure due to critical subsystem failure to that caused by subcritical system failure (i.e., find each one).

(d) If three of these systems each has one subsystem failure, what is the probability of at least one total system failure?

Statistical Measures

1. Given the following data set, compute the statistical measures, arithmetic and geometric means, median, and mode: 2.3, 5.8, 3.5, 6.3, 5.8, 2.3, 7.1, .5, 5.8.

2. Find the standard deviation of the above data set assuming, first, that the data are a sample and second, that it is a complete data set of a system.

3. Given the following data set with the respective probabilities for each value, compute the expected value and variance: $9.2E-3$ (.3), $2.5E-2$ (.2), $7.3E-3$ (.1), $1.5E-2$ (.3), $6.1E-4$ (.1).

Statistical Distributions

Binomial

1. A machine is producing parts that are found to be 8% defective. Calculate the probability of 0, 1, 2, 3, and more than 1 defective in samples of 8 parts taken from the production line. Assume stationarity and independence of the defectives.

2. Tests have determined that a relay has a mean failure probability of per cycle of $5.3E-3$.

(a) If the relay is cycled five times, what is the probability of at least one failure in the five cycles?

(b) If a system has 10 such relays installed and each relay is cycled 50 times in the course of system operation, what is the probability of at least 1 of the 10 relays having at least one failure in 50 cycles?

3. If an aircraft cargo door is closed but improperly locked, it is believed that there is a probability .003 of it opening in flight. What is the probability of any cargo door opening in flight in a year if 30 doors are improperly locked in a year?

4. An aircraft hydraulic system has three pumps. Two are needed to maintain system pressure for flight control operations. The failure probability of each of these pumps is found to be $3.5E-3$ per flight. If the aircraft in which these pumps are installed fly 2000 flights per year, what is the probability of complete flight control failure due to pump failure in a year of operations?

5. An important fire suppression system has three sensing units. Two operative units are needed for system function. Periodic inspections are made for inoperative sensors. Repeated inspections have disclosed that the probability of a sensor being inoperative is $4.8E-2$.

   (a) What is the probability of finding one inoperative sensor if a single sensor is inspected?

   (b) What is the probability of the system being inoperative?

   (c) What is the probability of finding the system inoperative if two units are inspected?

6. A space vehicle has five rocket thrusters to maintain proper orientation. Proper orientation will not be possible if two or more of the thrusters fail. Extensive testing has disclosed that 1% of the thrusters are subject to failure.

   (a) What is the probability of any one of the thrusters on one vehicle failing?

   (b) What is the probability of at least one thruster on a vehicle failing?

   (c) What is the probability of the vehicle orientation system failing due to thruster failures?

   (d) If four vehicles are to be launched in a program, what is the probability of at least one of the vehicles experiencing orientation failure due to thruster failure?

Normal Distribution

1. Measurements of the mean life of a video tube give a mean life of 1000 hours with a standard deviation of 175 hours. It is found that the life is normally distributed.

   (a) What part of these video tubes do we expect to have a life greater than 1325 hours?

   (b) What part of these video tubes do we expect to have a life less than 500 hours?

   (c) What part of these video tubes do we expect to have a life between 500 and 1325 hours?

   (d) What are the number of hours that one of these video tubes can be expected to function with a .95 probability that it will achieve that number of hours without a failure?

2. A particular model of hydraulic pump is found to have a mean time between failure of 800 hours with a standard deviation of 120 hours. The mean time between failure of this pump is normally distributed.

(a) What is the probability that this model of pump will fail between 630 and 950 hours?

(b) If it is desired to have a probability of pump failure at less than 600 hours be .10, what mean time to failure must be achieved by the population of pumps if the standard deviation remains 120 hours?

3. An examination of a radar system discloses a mean malfunction rate of 11 per month with a standard deviation of 2.5. The malfunction rate is normally distributed.

(a) What is the probability that there will be more than 15 malfunctions of this radar system in the next month?

(b) What are the number of malfunctions such that the radar system will have no less than a .07 probability of exceeding that number of malfunctions during the next month?

4. Tests of arresting cables produced by a certain company show that the mean number of arrestments to an unacceptable of strands failing is 1100. The standard deviation of unacceptable strands failing is 120. This parameter is normally distributed. If we wish to set the probability of unacceptable number of strands being failed at cable change at .003, what will be the maximum number of arrestments allowed prior to cable change?

Poisson

1. Testing of a number of switches finds that in 350,000 hours of total test time, 35 failures occur. It is assumed that the Poisson distribution will govern the probabilities of failure.

(a) What is the probability of a single switch operating for 1000 hours without a failure?

(b) What is the probability of a single switch experiencing at least one failure in 1000 hours?

(c) What is the probability of 6 of these switches experiencing more than 3 failures in a total exposure of 1000 hours per switch?

2. A system is tested for 12,500 hours and experiences three failures. Its failure rate is believed to be Poisson distributed.

(a) What is the probability of no failures in 5000 hours of operation of this system?

(b) If two of these systems are each operated 1500 hours, what is the probability of any failures in the two systems?

(c) If the desired probability of any failures in two systems in 1500 hours is .03, what must the true mean failure rate of each system be to achieve this desired standard?

3. The error rate of a radar controller handling a specified sector is found to be .002 per target. The probability of the controller is assumed to be governed by the Poisson distribution.

(a)   What are the maximum number of of targets the controller may handle in a shift if the probability of any errors is to be no greater than .06?

(b)   If twice the number of targets as determined in (a) above are expected to be handled by the controller on a shift, and a second controller is placed on the shift, what are the maximum number of targets that may be handled by both controllers to maintain the probability of any errors at .06?

Chi-Square: Confidence Bounding

1. In a replacement test of a component, 15 items are each tested for 1000 hours. During this test seven failures occur.

(a)   What is the 90% upper confidence bound on the component failure rate?

(b)   What is the 75% lower confidence bound on the failure rate?

(c)   What is the 75% lower confidence bound on the probability one of these components surviving 1000 hours of service.

2. A contract is written specifying the failure rate of a space system locking mechanism of $8.0E-4$ with 90% confidence.

(a)   If 6000 locking attempts are funded in the test program, can the rate be proved with the desired confidence if it is assumed that two failures will occur in 6000 attempts? What is the failure rate that can be proved with the required confidence?

(b)   How many failures can be accepted if the failure rate is to be proved with the desired degree of confidence?

(c)   If there is the most desirable outcome to the test program and the entire 6000 funded tests are completed, what is the rate that may be proved with 90% confidence?

3. A fire-warning sensing mechanism is tested and fails to activate properly three times in 500 tests.

(a)   What is the 90% conservative, confidence bounded rate that is proved by this test?

(b)   If the desired 90% confidence bounded failure rate is $1.1E-2$, how many additional tests must be made to prove the desired rate if no additional failures occur?

(c)   What is the confidence that has been proved in the desired rate with the test as performed (i.e., 500 tests, three failures).

4. A contract is written specifying that an accident rate of $6.8E-6$ per hour will be proved with a confidence of .9 during a system test. The test will be funded to a level that will allow two accidents during test while proving the desired rate.

(a) What is the required budget for the test if the testing will cost $100 per hour?

(b) Given that the budget is granted and used for testing, what is the maximum confidence that can be hoped for from system test?

(c) Given that the budget is spent, what is the most favorable rate that can be proved with 90% confidence?

(d) Given that no failures occur during the test, how many dollars may be saved in the test program by terminating the test at the earliest opportunity with proof of the rate with the required confidence?

REFERENCES AND BIBLIOGRAPHY

Ben-Horim, M. and Levy, H., *Statistics: Decisions and Applications in Business and Economics*, 2nd ed. Random House, New York, 1984.

Bohrnstedt, G.W. and Knoke, D., *Statistics for Social Data Analysis*. F.E. Peacock, Itasca, Illinois, 1982.

Draper, N.R. and Smith, H., *Applied Regression Analysis*, Wiley, New York, 1966.

Epstein, B., "Estimation from Life Test Data," *IEEE Transactions on Reliability and Quality Control*, Vol RQC-9, April 1960.

Fischer, R.A., *Statistical Methods and Scientific Inference*. Oliver Boyd, London, 1956.

Freund, J.E., *Statistics: A First Course*. Prentice-Hall, Inc., Englewood Cliffs, NJ, 1981.

Hines, W.H. and Montgomery, D.C., *Probability and Statistics in Engineering and Management Science*, 2nd ed., Wiley, New York, 1980.

Hocking, R.R., "The Analysis and Selection of Variables in Linear Regression," *Biometrics*, Vol. 32, pp. 1–49, 1976.

Montgomery, D.C., *Design and Analysis of Experiments*, Wiley, New York, 1976.

Parsons, R., *Statistical Analysis: A Decision Making Approach*. Harper & Row, New York, 1978.

Shooman, M.L., *Probabilistic Reliability: An Engineering Approach*. McGraw-Hill, New York, 1968.

Winkler, R.L. and Hays, W.L., *Statistics: Probability, Inference and Decision*, 2nd ed. Holt, Rinehart and Winston, New York, 1975.

NETWORK ANALYSIS

18

EVENT SYSTEMS

There are several methods by which events occurring within a time structure may be organized to describe a particular aspect of system function. Program Evaluation Review Techniques (PERT) networks are a way of organizing events in a construction or production process to facilitate scheduling and planning of that process. A fault tree is a network method that is used to analyze specified outcomes of system function for the cause of these outcomes. Networks may be used to examine desired or undesired outcomes. A fault tree is generally used to examine an undesired outcome of system function, although it may also be used for desired results. Network analysis may be quantitative or qualitative. Qualitative analysis may result in critical event specification or a nonquantitative measure of system state. Quantitative analysis will provide numerical measures of system performance such as probabilities or dollars return.

A series–parallel network is a generalized type of network organization of which the above two types of networks are subsets. Generally and traditionally, series–parallel networks are associated with *reliability systems*. In this capacity, they model *reliable* functioning of a system or a desired outcome. Probabilistic numerical results of reliability system analysis will be the probability of the system functioning reliably. It has probabilistic outputs that are large.

We may also organize sets of events in a series–parallel system in such a way that their successfully functioning will result in an *accident*. Such an organization of events is called an *accident system*. It is common to think of the word successful as describing something good or desirable. In the case of network analysis this is not always true. The word "successful" in this context merely means that the network functions. If the functioning network leads to proper system function, then it is good and we may speak of the system reliably functioning. If the specified network leads to an accident or an undesired event of some other sort, then successful functioning of the network is *not* good. We may then speak of such a network as an accident system.

The focus in this text is on accident systems. Because of the unfamiliarity of such a network, we will approach them from the point-of-view of reliability systems.

The output of an accident system may be described by a probability value. It will also have attached to it a measure of accident severity. Thus, the output of an accident system may be thought of as probability density functions of occurrence and severity. In this chapter, we will examine the probabilistic nature of series–parallel networks. Severity will be addressed in later chapters.

A series–parallel accident system can be composed of normal component events, component failure events, human normal and abnormal events, as well as environmental events. The probabilities of these events may range across the probability scale from 0.00 to 1.00. Normal events would be expected to have probabilities approaching 1.00. Failure events would, hopefully, have probabilities near to 0.00.

We will gain understanding by contrasting reliability and accident systems. In general, the accident outcome of a system will not be the complement of that system treated as a reliability system.

18.1 SERIES SYSTEMS

Let us consider a series system first as a reliability system, then as an accident system. We will examine these systems in their simplest form and then expand to a system a bit more complex.

Figure 18.1 depicts a series system consisting of two events, 1 and 2. The probability of event 1 is .9 and event 2 is .8, as can be seen in the figure. The logic of a series system will be assumed to mean merely that all events of the system must occur for the system to function. The order of the event occurrence is not material. It is only necessary that all events occur in the time constraint t.

The probability of successful or reliable functioning of the above system will be:

$$P(s) = P(1)P(2) = (.91)(.92) = .8372 \qquad (18.1)$$

It is apparent that such a system does not make a desired arrangement for a reliability system. For instance, all events in the system must function for the system to be functionally reliable. Also, examining the $P(s)$, it can be seen that it is a smaller number than either of the events that make up the system, therefore, it is less desirable as a reliability number. However, if we consider a series system as an accident system, the outlook will be quite different.

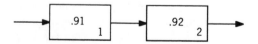

Figure 18.1 Series reliability system.

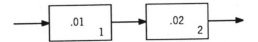

Figure 18.2 Series accident system.

Figure 18.2 depicts a series accident system. Note that although the logic of the system remains the same as a reliability system, the probabilities of the events are small as they must be in an accident system of reasonable safety.

$$P(\text{acc}) \; = \; P(1)P(2) \; = \; (.01)(.02) \; = \; 2E{-}4 \qquad (18.2)$$

One can see at once, that this system logic structure is desirable as an accident system. Both or all events of the system must occur to cause the accident. In addition, the probability of system function is less than the probabilities of the events in the system. Such a system structure allows one to prevent the accident by ensuring that one or more of the events in the series does *not* occur.

As series systems become larger (see Figure 18.3), their desirable and undesirable characteristics increase accordingly. For instance, a series reliability system of 300 events each with a probability of reliably functioning over the time constraint of .99, would be quite unreliable.

$$P(s) \; = \; (.99)^{300} \; = \; .0490$$

On the other hand, a series accident system containing a large number of events each with a rather large probability of occurring over the time constraint, will be quite safe (see Figure 18.4).

$$P(\text{acc}) \; = \; (.8)^{300} \; = \; 8.453E{-}30$$

An example of a series system would be Figure 18.5. As a reliability system it can be seen that all of the components must function reliably if the system is to function.

The probability of the system functioning reliably is the product of the reliabilities of each of the components.

$$R \; = \; P(s) \; = \; (.99)(.85)(.91)(.94) \; = \; .7198$$

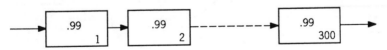

Figure 18.3 Large series reliability system.

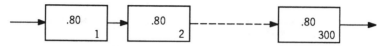

Figure 18.4 Large series accident system.

18.2 PARALLEL SYSTEMS

A parallel system is often referred to as a redundant system. Such a system will function reliably if fewer than all of the components or events that make up the system function. Figure 18.6 depicts a simple one-of-two system.

The advantage of a parallel system is in its increased probability of successful functioning over that of the probabilities of the events that make up the system. In the foregoing one-of-two system, either one of the events will suffice for system success or reliability.

$$R = P(s) = P(\geq 1) = 1 - P(0) = 1 - (.1)(.2)$$

$$= 1 - .02 = .98 \tag{18.3}$$

The calculation in equation (18.3) finds the probability of at least one event functioning by calculating the complement of both events failing. This is obviously a most desirable system logic structure for a reliability system.

In the case of an accident system, a parallel system logic is undesirable. Figure 18.7 depicts a parallel accident system.

In the one-of-two parallel accident system, we can see that either event can cause function flow and the accident. This in itself should cause alarm. In effect, we are looking at two single point failures. The probability of an accident may be calculated as in the case of a reliability system by calculating the probability of at least one event occurring.

$$P(\text{acc}) = P(\geq 1) = 1 - P(0) = 1 - (.99)(.98) = 1 - .9702$$

$$= .0298 \tag{18.5}$$

Note that the probability of an accident has increased over that of the probability of either of the two events that make up the system. An accident system with a large number of parallel events would almost ensure the accident occurrence.

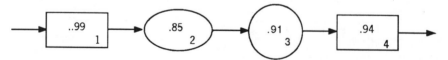

Figure 18.5 Series cooling system.

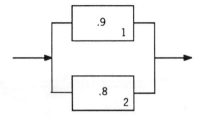

Figure 18.6 Parallel reliability system.

The probability of an accident in the case of the system of Figure 18.8 is:

$$P(\text{acc}) = 1 - P(0) = 1 - (.99)^{500} = .9934 \qquad (18.5)$$

Figure 18.9, depicts a parallel reliability system.

One of more of the pumps will provide the needed flow for system reliable functioning. If the probability of each pump functioning successfully for the required time period is .9, then the reliability of the system may be calculated as follows:

$$R = P(s) = P(\geq 1) = 1 - P(0) = 1 - (.1)^3 = .9990 \qquad (18.6)$$

18.3   SERIES–PARALLEL SYSTEMS

A more complex system that consists of both series and parallel logic structures can be analyzed by using the rules of both series and parallel networks. Figure 18.10 is an example of such a network.

Rather than provide simple probabilities for the events of this system, mean times to event occurrence (m) are given. This will illustrate that all probabilities are governed by event occurrence rates and an exposure or time constraint (t). This will also facilitate the use of this network as an accident or reliability system.

First, treating the network of Figure 18.10 as a reliability system it would be helpful, if not essential, to write a master equation for the solution of this system. Equation (18.7) is this solution.

$$R = P(s) = (1 - P(f_1)P(f_2)P(f_3))(1 - P(f_4)P(f_5)) \qquad (18.7)$$

Equation (18.7) states that the probability of system success is the probability of one or more successes in the first set of three events and the probability of one

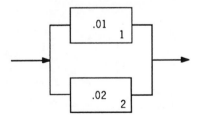

Figure 18.7 Parallel accident system.

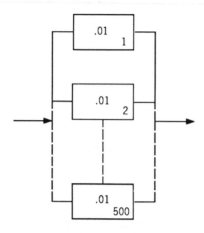

Figure 18.8 Multiple parallel accident system.

or more successes in the second block of two events. The equation treats the system as a set of two series subsets with the probability of at least one path through each subset as the system solution. Assuming a system exposure time of 300 hours, the system may be solved as follows noting that:

$$P(f) = 1 - e^{-\lambda t} = 1 - e^{-t/m} \tag{18.8}$$

Therefore:

$$P(f_1) = 1 - e^{-300/2.3E3} = .1223$$

Finding other probabilities of failure similarly, the solution for the reliability of the network is as follows:

$$P(s) = [1 - (1223)(.0645)(.6575)][1 - (.0343)(.0142)]$$

$$= .9943$$

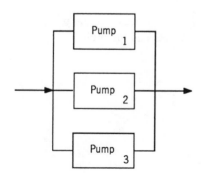

Figure 18.9 Parallel pumping system.

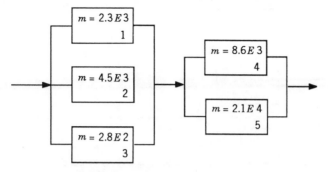

Figure 18.10 Series-parallel system.

Next, treating this system as an accident system, quite a different solution would be forthcoming. We will make the simplifying assumption that the accident system will consist of events logically arranged as in Figure 18.10. The rather large MTBFs (*m*) of the system events will also be appropriate for events in accident causation paths.

$$P(\text{acc}) = (1 - P(O_1)P(O_2)P(O_3))(1 - P(O_4)P(O_5)) \qquad (18.9)$$

where $P(O_i)$ is the probability that no events will occur in the time constraint or system exposure.

Equation (18.9) states that an accident will occur if at least one event occurs in the first and second system blocks. Noting that:

$$P(O_i) = e^{-\lambda_i t}$$

$$P(\text{acc}) = (1 - (.8777)(.9355)(.3425))(1 - (.9657)(.9858))$$

$$= 3.451E{-}2 \qquad (18.10)$$

Although the probability of an accident is appropriately small, it should be observed that it is not the complement of reliability. This is because the use of the same system logic structure for both accident and reliability systems causes one to be a partial *dual* of the other. Although we will consider system duals in greater detail under fault tree analysis, we should observe here that a system dual is one that has complementary logic. In such logic, series relations become parallel and parallel become series. Thus, in the second block of two events, the reliability system requires at least one event to function reliably. This is a parallel functional relationship. In the accident system however, both events must fail. This is a series functional relationship.

Some networks may not be in a series–parallel relationship initially but can be altered to be in such a relationship by a few simple construction lines. Consider Figure 18.11.

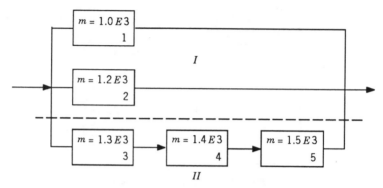

Figure 18.11 Partitioned system.

The system of Figure 18.11 was not in series–parallel logic as originally structured. However, the addition of a simple construction line, as shown in the figure, partitioning the system into two parallel subsystems (I and II) each of which are in a series–parallel logic structure, will facilitate the system solution.

The partitioned subsystems take the following form (Figure 18.12).

Addressing this partitioned system as an accident system, the following solution equation may be written.

$$P(\text{acc}) = 1 - P(O_\text{I})P(O_\text{II}) \tag{18.11}$$

Having written the master solution equation for this system, it is necessary to solve for its elements, assuming a time constraint of 250 hours.

$$P(O_\text{I}) = 1 - P(O_1)P(O_2) = 1 - (.7788)(.8119) = .3677$$

$$P(O_\text{II}) = 1 - [(1 - P(O_3))(1 - P(O_4))(1 - P(O_5))]$$

$$= 1 - ((.1750)(.1635)(.1535)) = .9956$$

The final solution of equation (18.11) would then be as follows:

$$P(\text{acc}) = 1 - (.3677)(.9956) = .6339$$

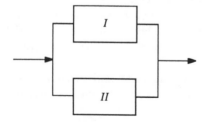

Figure 18.12 Partitioned subsystems.

This rather large probability for an undesired event is due not so much to the probabilities of the individual events that are reasonably small but to the logic of the system. Note that there are two single-point failures, either of which could cause the accident.

Partitioning complex systems into series-parallel subsystems, will not always allow a series-parallel solution to the system. For a general solution we must look to other methods.

19

BOOLEAN ALGEBRA

Prior to investigating other forms of network analysis, we will need an understanding of a mathematical tool called Boolean algebra. We will not examine all of the aspects of Boolean algebra nor the wide variety of problems that may be treated by this unique form of algebra. In this brief examination of the subject, we will merely set forth the elements of Boolean algebra that will enable us to understand the fault tree logic process, and the method by which we will reduce a fault tree to its minimal cut sets.

19.1 BOOLEAN OPERATORS

As an aid in assisting in the understanding the principles of Boolean algebra, we should examine Venn diagrams. Venn diagrams are a method of depicting sets of events and their probabilities as geometric space. Such a body of knowledge is part of set theory.

A set is simply a collection of events with particular characteristics. A subset is a collection of events within the primary set with further defined characteristics. We might define a set as all accidents involving a particular system that occurred in a defined period of time. A subset (A) might be accidents in which fatal injuries occurred. Another subset (B) might be accidents that had no injuries. Finally, a third subset (C) might be accidents of the parent set in which the operator was under the influence of drugs.

Figure 19.1 is a Venn diagram of the set and subsets of this situation.

In the foregoing Venn diagram the universe, as defined by the rectangle, bounds the parent set of all elements of this system. The other sets are defined by a labeled space representing all of the events of a particular characteristic. The null or empty set is labeled with the notation ∅.

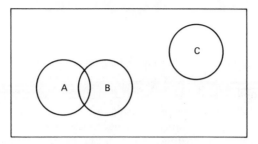

Figure 19.1   Venn diagram.

ω = The universe

A = Events of a particular characteristic

B = Events of another characteristic, some involving A

C = Events of another characteristic

We can see from the Venn diagram that sets *A* and *B* have a common space or characteristic. Set *C* has no common characteristic with either of the other two sets. In addition there remains some part of the universe not defined by the subsets.

There are notations that may help us describe combinations of these events.

$$A \cap B \qquad\qquad (19.1)$$

The notation of equation (19.1) stands for a Boolean *intersection*. The notation reads *A* intersect *B*. An intersection is a characteristic of two or more subsets. For instance, in studying drug-related accidents it could be a subset that was drug related and fatal.

$$A \cup B \qquad\qquad (19.2)$$

The notation of equation (19.2) stands for Boolean *union*. The notation reads *A* union *B*. A union is a combination of characteristics in two or more subsets. For instance, in the study of drug related accidents it could be the characteristic of having either a drug-related accident or a fatal accident. The difference between union and intersection in this example is that the union includes drug related accidents that are not fatal and fatal accidents that are not drug related. However, a union would not consider any events twice.

$$\overline{A} \qquad\qquad (19.3)$$

The notation of equation (19.3) means *not A* or, if a probability is associated with $\overline{A}$, the complement of *A*. For instance, if *A* is a fatal accident, then $\overline{A}$ is nonfatal accident.

$$P(A|B) \qquad\qquad (19.4)$$

The notation of equation (19.4) means a conditional probability of A given that B has occurred. For instance, one may be concerned with the probability of a drug-involved accident given that a fatal accident has occurred. To be more precise, conditional probability is the probability of A events taken from the B subset of events. Figure 19.2 depicts this relationship.

To calculate such a probability, it is necessary to only count the number of events that satisfy the specifications of both A and B and divide that number by the number of events in B. In terms of our previously defined notations:

$$P(A|B) = \frac{P(A \cap B)}{P(B)} \qquad\qquad (19.5)$$

Equation (19.5) is a general expression for conditional probability. Under special circumstances this equation may be simplified.

Let us assume that in the foregoing Venn diagram:

$$A = 20 \text{ events}$$

$$B = 30 \text{ events}$$

$$C = 10 \text{ events}$$

$$A \cap B = 5 \text{ events}$$

$$\omega = 120 \text{ events}$$

From a simple count of the number events that conform to the specifications of A, B, C, and the universe, we may make calculations.

$$P(A) = 20/120 = .1667$$

$$P(B) = 30/120 = .2500$$

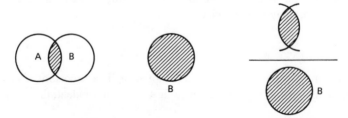

Figure 19.2   Conditional probability of A given B.

$$P(C) \; = \; 10/120 \; = \; .0833$$

$$A \cup B \; = \; 20 \; + \; 30 \; - \; 5 \; = \; 45 \text{ events}$$

$$P(A \cup B) \; = \; 45/120 \; = \; .3750$$

$$A \cup C \; = \; 20 \; + \; 10 \; = \; 30 \text{ events}$$

$$P(A \cup C) \; = \; 30/120 \; = \; .2500$$

$$P(A|B) \; = \; 5/30 \; = \; .1667$$

Note that $P(A|B) = P(A)$. This should suggest that the occurrence of event B in no way influences the occurrence of event A. This is the fundamental meaning of independence.

If two subsets are related such that knowledge of an event that satisfies the argument for both subsets in no way influences the probabilities of occurrence of events in either of the subsets, then these events must be independent. Thus, from the foregoing calculations:

$$P(A|B) = P(A) \tag{19.6}$$

Likewise, two events are independent if $P(A \cap B) = P(A)P(B)$. Carrying this relation through:

$$P(A|B) \; = \; \frac{P(A \cap B)}{P(B)} \; = \; \frac{P(A)P(B)}{P(B)} \; = \; P(A) \tag{19.7}$$

Another characteristic that is useful in describing sets is *mutually exclusive*. Two subsets are mutually exclusive if an event satisfying the argument for one subset precludes satisfaction of the argument of the other subset. In the Venn diagram, sets A and C are mutually exclusive as are sets B and C. If events are mutually exclusive their union is the sum of the events or their probabilities. This can be seen in the above calculations. Another term for mutually exclusive is *disjoint*.

We may then write the general expression for the union of two subsets.

$$P(A \cup B) = P(A) + P(B) - P(A \cap B) \tag{19.8}$$

If the events are mutually exclusive, the unions become the simple sum of the events or their probabilities.

Recall that in the section on probability it was stated that we could find the probability of either of two types of events occurring in a single exposure by adding their probabilities if the events were mutually exclusive. Likewise, we said that we could find the probability of two types of events occurring in two exposures by multiplication if the events were independent. These restrictions can be observed in the above calculations.

There remains the question of whether two subsets being mutually exclusive precludes their being independent. The answer is yes, except in the trivial case of the primary subset being empty or a null set.

The commutative, associative, and distributive laws also apply in Boolean algebra.

$$\text{Commutative Laws: } A \cup B = B \cup A \qquad (19.9)$$

$$A \cap B = B \cap A$$

$$\text{Associative Laws: } (A \cup B) \cup C = A \cup (B \cup C) \qquad (19.10)$$

$$(A \cap B) \cap C = A \cap (B \cap C)$$

$$\text{Distributive Laws: } A \cup (B \cap C) = (A \cup B) \cap (A \cup C) \qquad (19.11)$$

$$A \cap (B \cup C) = (A \cap B) \cup (A \cap C)$$

A practical example of the application of Boolean algebra would be helpful. Let us assume that, in an industrial situation, we have a problem with certain workers who have bronchial problems. Some of these workers are working in the paint shop and there is concern as to whether this condition is being caused by breathing toxic substances in the paint shop. There is also concern as to whether age of the workers is another factor in causation.

We define three subsets and their intersections. Then it is a simple matter to go to the shop floor and count workers in the subsets.

A = Workers in the paint shop = 75
B = Workers over 40 years of age = 150
C = Workers with bronchial problems = 25
ω = Total workers = 250
$A \cap B \cap C = 18$
$A \cap B = 30$
$A \cap C = 21$
$B \cap C = 20$

The Venn diagram of this situation is shown in Figure 19.3.

Finding the probability of bronchial problems given that the worker is in the paint shop:

$$P(C|A) = 21/75 = .2800$$

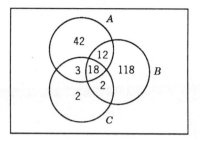

Figure 19.3   Venn diagram of bronchial problems.

To determine if subsets C and A are independent:

$$P(C) = 25/250 = .1000$$

The subsets are not independent because the $P(C|A)$ does not equal $P(C)$. The probability of having bronchial problems given work in the paint shop is higher than the probability of bronchial problems for the entire work force. However, before we reach the conclusion that the paint shop is the cause we might investigate the age relationship. Age could be the cause and older workers might be disproportionately assigned to the paint shop.

Examining the effect of age on bronchial problems:

$$P(C|B) = 20/150 = .1333$$

Subsets C and B are not independent because the probability of the older workers having the bronchial condition is also higher than for the entire work force.

Finally, examining the tendency for older workers to be assigned to the paint shop:

$$P(A|B) = 30/150 = .20$$

Investigating the independence of subsets A and B:

$$P(A) = 75/250 = .3000$$

We see that, in fact, there is a tendency for younger workers to be assigned to the paint shop. Therefore, barring other factors that have not been investigated, we can conclude that there is a condition in the paint shop that is apparently causing bronchial problems.

To provide additional practice in visualizing probability spaces through Venn diagrams, the following examples are offered (see Figure 19.4).

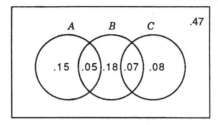

Figure 19.4   Venn diagram for Boolean problems.

1. $P(A \cup B) = .45$

2. $P(A \cup C) = .35$

3. $P(\overline{A} \cap C) = .15$

4. $P(\overline{A} \cap B) = .25$

5. $P(\overline{A} \cap \overline{C}) = .65$

6. $P(\overline{A} \cap \overline{B}) = .55$

7. $P(\overline{A} \cup B) = .85$

8. $P(A \cup \overline{B}) = .75$

9. $P(\overline{A} \cup \overline{B}) = .95$

10. $P(\overline{A} \cup \overline{C}) = 1.00$

11. $P[(A \cap B) \cap C] = 0.0$

12. $P[(\overline{A} \cap C) \cap C] = .15$

13. $P[(A \cap \overline{C}) \cup C] = .35$

14. $P[(\overline{A} \cap C) \cup \overline{C}] = 1.00$

15. $P[(A \cap B) \cap \overline{C}] = .05$

16. $P[(A \cup B) \cap C] = .07$

There are a few Boolean identities that are of some importance. These are Boolean equations that stand true for any Venn diagram.

$$P(\omega) = 1.00$$

$$P(\phi) = 0.00$$

$$P(A \cup \omega) = 1.00$$

$$P(A \cap \omega) = P(A)$$

$$P(A \cup A) = P(A)$$

$$P(A \cap A) = P(A)$$

19.2   CONDITIONAL PROBABILITY—BAYES' THEOREM

We will briefly address Bayes' theorem because of its extension of conditional probability. Consider the Venn diagram of Figure 19.5.

Figure 19.5 depicts subsets B_1, B_2, B_3, and B_4 that are mutually exclusive and exhaustive. Subset A intersects the B events. Event A may is seen to be the sum of the intersections with B_i.

$$P(A) = P(A \cap B_1) + P(A \cap B_2) + P(A \cap B_3) + P(A \cap B_4) \quad (19.12)$$

From equation (19.5) we can observe:

$$P(A \cap B_i) = P(A|B_i)P(B_i) \tag{19.13}$$

Substituting into equation (19.12):

$$P(A) = P(A|B_1)P(B_1) + P(A|B_2)P(B_2) \tag{19.14}$$
$$+ P(A|B_3)P(B_3) + P(A|B_4)P(B_4)$$

Now suppose it is desired to calculate a specific conditional probability of one of the B_i subsets given A, specifically $P(B_3|A)$. We have previously shown that:

$$P(B_3|A) = \frac{P(A \cap B_3)}{P(A)} \tag{19.15}$$

Therefore:

$$P(B_3|A) = \frac{P(A|B_3)P(B_3)}{P(A|B_1)P(B_1) + P(A|B_2)P(B_2) + P(A|B_3)P(B_3)P(A|B_4)P(B_4)}$$

$$\tag{19.16}$$

The generalized equation for Bayes' theorem is:

$$P(B_i|A) = \frac{P(A|B_i)P(B_i)}{\Sigma P(A|B_i)P(B_i)} \tag{19.17}$$

A practical problem will illustrate the method.

Let us examine three classes of smokers, heavy, light, and nonsmokers. We will further assume that the national statistics show that:

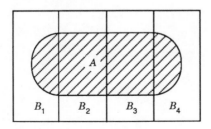

Figure 19.5 Venn diagram illustrating Bayes theorem.

$$P(\text{nonsmoker}, B_1) = .60$$

$$P(\text{light smoker}, B_2) = .25$$

$$P(\text{heavy smoker}, B_3) = .15$$

$$P(\text{cancer}|\text{nonsmoker}, P(A|B_1)) = 1E-6 \text{ (per life)}$$

$$P(\text{cancer}|\text{light smoker}, P(A|B_2)) = 8E-5$$

$$P(\text{cancer}|\text{heavy smoker}, P(A|B_3)) = 3E-4$$

Calculating the conditional probabilities of being nonsmoker, light smoker, or heavy smoker given that you develop lung cancer:

$$P(B_1|A) = \frac{P(A|B_1)P(B_1)}{\Sigma P(A|B_i)P(B_i)} = \frac{6E-7}{6.56E-5} = .0091$$

$$P(B_2|A) = \frac{P(A|B_2)P(B_2)}{\Sigma P(A|B_i)P(B_i)} = \frac{2E-5}{6.56E-5} = .3049$$

$$P(B_3|A) = \frac{P(A|B_3)P(B_3)}{\Sigma P(A|B_i)P(B_i)} = \frac{4.5E-5}{6.56E-5} = .6860$$

20

CUT SETS

There are many systems with a logic that cannot be partitioned into subsets or series–parallel relationships. Therefore, a general analytical method is needed that will treat all system networks, no matter how complex. The method of cut sets is such an approach. This method can be investigated more fully in reliability literature.

A cut set is set of events in a logic-linked system of nondependent events, that will assure the system will not function. If the system in question is a reliability system, the occurrence of all events in a cut set will assure a state of unreliability. Of course, cut sets applied to an accident series–parallel diagram will prevent the accident. Our use of cut sets will be to find sets that will cause the upper or top event in a fault tree. We will continue to call them cut sets from their reliability origins although, of course, they will be logical path sets of a fault tree.

We will make use of some Boolean algebra concepts to develop the basis for cut set theory.

Figure 20.1 is a Venn diagram of three interlocking subsets in a Boolean universe.

In this particular diagram we may state:

$$P(A + B + C) = P(A) + P(B) + P(C) - P(AB) - P(AC)$$

$$- P(BC) + P(ABC) \qquad (20.1)$$

In general, if we have subsets, $A_1, A_2, \cdots A_i$ and do not know the amount of intersection between them, we cannot make an exact statement about $P(A_1 + A_2 + \cdots A_i)$. However, noting the nature of the higher order terms in the case of Figure 20.1, we may always state:

$$P(A_1 + A_2 + \cdots A_i) = \leq P(A_1) + P(A_2) + \cdots P(A_i) \qquad (20.2)$$

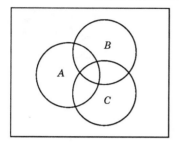

Figure 20.1   Venn diagram of Boolean spaces.

As the probabilities of the cut sets A_i become smaller, equation (20.2) approaches equality. If the subsets are mutually exclusive then equation (20.2) becomes an exact equation.

If the system we are examining leads to a desired system state, the sum of the probabilities of the cut sets is a conservative upper bound for the probability of at least one of the cut sets entering a fault state and thus causing the undesired or accident event. It then remains to find exhaustive and nonredundant sets of events each of which may cause an accident event. The nonredundant requirement for these sets merely means that no cut set is included within any other cut set. We will call nonredundant cut sets *minimal* cut sets.

20.1   CUT SET DEVELOPMENT

Figure 20.2 shows a network diagram of a very simple reliability system.

In this network, we have two paths through the system, 1, 2 and 3. Cutting these paths at any point will ensure that the system does not function.

The cut sets of the above system may be found by inspection as the following:

$$C_1 = 1, 3$$

$$C_2 = 2, 3$$

$$C_3 = 1, 2, 3$$

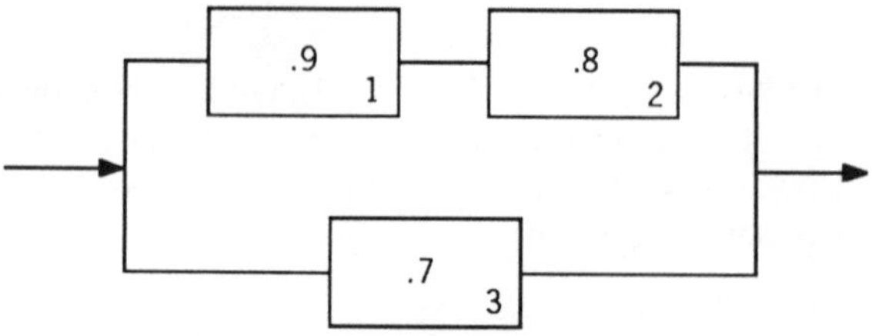

Figure 20.2   System network.

If we examine the above three cut sets, we can see that one, cut set 3, is redundant or nonminimal. Cut set 3 is a cut set that contains at least one of the other cut sets. The minimal cut sets of this system are then:

$$C_1 = 1, 2$$

$$C_2 = 2, 3$$

Since each of these minimal cut sets will cut the system, the upper bound on the probability of system failure of this reliability system is:

$$P(f_s) = P(C_1) + P(C_2) \tag{20.3}$$

The probability of C_1 occurring is the probability of event 1 and event 2 failing:

$$P(C_1) = (.1)(.3) = .03$$

The probability of C_2 occurring is found in a similar manner:

$$P(C_2) = (.2)(.3) = .06$$

Thus, the upper bound on the probability of system failure is:

$$P(f) \leqslant .03 + .06 = .09$$

A more complex example will illustrate the method. Figure 20.3 depicts a command signal system.
A reliability block diagram of the system takes the form of Figure 20.4. Probabilities shown are event reliabilities.

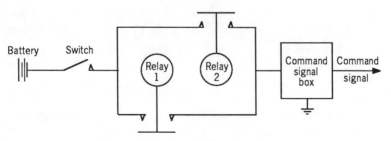

Figure 20.3 Command signal system.

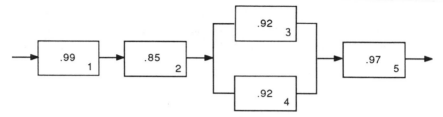

Figure 20.4 Reliability block diagram of command signal system.

Event 1 = Battery power

Event 2 = Switch closed

Event 3 = Relay 1 closed

Event 4 = Relay 2 closed

Event 5 = Command signal box functions

The cut sets of the above system are the following:

$$C_1 = 1 \quad C_3 = 3, 4$$
$$C_2 = 2 \quad C_4 = 5$$

The probabilities shown for the events are reliabilities or the probabilities that the event will occur in the time constraint t.

The upper bound on the probability of system failure is:

$$P(f_s) \le .01 + .15 + (.08)(.08) + .03 = .1964$$

Cut sets will always solve the system no matter how complex the logic.

20.2 SYSTEM DUALS—PATH SETS

Path sets are sets of events that assure system functioning. Thus, path sets must each cut all cut sets. In the case of the foregoing example, the path sets are:

$$P_1 = \overline{1}, \overline{2}, \overline{3}, \overline{5}$$

$$P_2 = \overline{1}, \overline{2}, \overline{4}, \overline{5}$$

It is apparent that events 1, 2, and 5 must be in each path set. These paths sets may be confirmed by examining Figure 20.3. It is clear that if there is battery power, switch closed, either relay 1 or 2 closed, and the command signal box functioning that the system will function.

A more structured solution for path sets in complex systems is through the system dual. A system dual is a system that has complementary logic of the original system and events in the dual are complements of the original system events. In a simple reliability block diagram, series relations become parallel and parallel relations become series. Figure 20.5 is a dual of Figure 20.4.

Having created the dual of the original system, we can find the cut sets of the dual. These cut sets will be the path sets of the original system.

In Figure 20.5 it can be seen that the cut sets are:

$$C_1 = \bar{1}, \bar{2}, \bar{3}, \bar{5}$$

$$C_2 = \bar{1}, \bar{2}, \bar{4}, \bar{5}$$

These are, of course, the path sets of the original system.

If we can imagine a system in which the events of the reliability system are identically arranged for the accident system, we will find the essence of the relationship between accident and reliability systems as well as the relationship between cut sets and path sets.

If we take the system of Figure 20.3 and imagine that an accident will occur if the command signal is not output during the time constraint, we will have such a relationship. The dual of that system becomes an accident system. The cut sets of the original system become the path sets of the dual and thus the sets that will cause an accident. The cut sets of the dual of the accident system will be the sets that prevent the accident. This can be confirmed when we observe that the cut sets of the dual are also the path sets of the reliability system.

When we examine fault trees we will observe these relations will continue to exist between cut sets and path sets of the original tree and its dual. In the case of the fault tree, the system logic is presented in its accident system logic. Cut sets as found on the fault tree will be logical paths to the Top event and it will be the path sets that prevent the occurrence of the Top event.

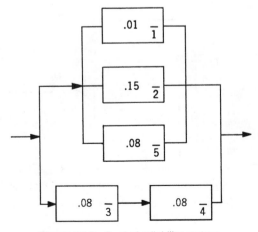

Figure 20.5 Dual of reliability system.

EXERCISES FOR PART III

1. Define the difference between an accident system and a reliability system.

2. What does an event in the dual of a reliability system represent?

3. Given the following system diagram with the mean time between events, m, and an exposure time $t = 125$:
   (a) Compute the probability of system failure assuming the system is a reliability system by the series–parallel method.
   (b) Compute the probability of an accident assuming the system is an accident system by the series–parallel method.

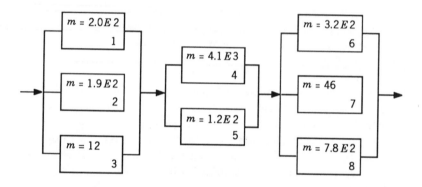

4. Given the following accident system with the mean time between event occurrence, m, as shown, compute the probability of an accident by the series–parallel method. Assume that $t = 85$.

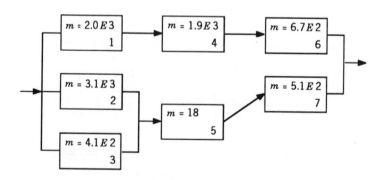

5. Given the following reliability system with the probabilities of reliable functioning of the events as shown, compute the upper bound on the probability of system failure by the cut set method.

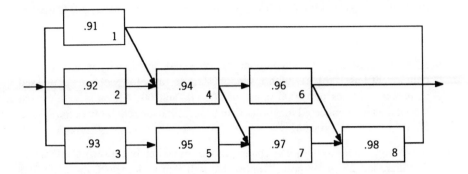

6. Given the following Venn diagram with the probabilities of events A, B, and C as shown, compute the probabilities indicated below by Boolean methods.

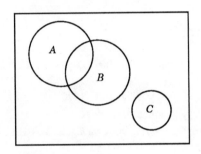

$P(A) = .5$

$P(B) = .2$

$P(C) = .1$

Events A and B are independent.

(a) $P(A \cap B) =$

(b) $P(A \cap C) =$

(c) $P(B \cap \overline{C}) =$

(d) $P(A \cup C) =$

(e) $P(A \cup B) =$

(f) $P(A \cap \overline{B}) =$

(g) $P(B \cup C) =$

(h) $P(B \cup \overline{C}) =$

(i) $P[(A \cap B) \cap \overline{C}) =$

(j) $P[(A \cap B) \cup \overline{C}) =$

(k) $P[(A \cap C) \cup B] =$

(l) $P[(B \cup C) \cap \overline{A}) =$

(m) $P[(A \cap \overline{B}) \cap C] =$

(n) $P[(B \cap C) \cup A] =$

(o) $P[(A \cap A) \cap B] =$

(p) $P[(A \cap \overline{B}) \cap B] =$

7. Consider the relative safety of the twin-engine aircraft versus the single-engine aircraft. Assume that the probability of engine failure in both installations is .005 per mission.

   (a) What is the probability of each aircraft type surviving a mission?

   (b) What is the percentage improvement of the twin-engine aircraft over the single engine?

8. A system consists of three major components. Component 1 has 50 parts, each with a probability of survival of mission ($P(s)$) of .98. Component 2 has 75 parts each with a $P(s)$ of .97. Component 3 has 100 parts each with a $P(s)$

of .997. All parts within each component are connected in series. Calculate the system probability of survival (reliability) if components 2 and 3 are functionally in parallel and component 1 is in series with components 2 and 3.

9. A system consists of three components connected functionally in series. Each component has 100 parts with respective failure rates for each part expressed in percent failures per 1000 hours of 20, 40, and 60. All parts within each component are connected functionally in series. Find the reliability of the system after 25 hours of exposure.

10. A weapon-fusing system can change mode immediately following release so that there are 25 single-point failures that will arm it. If armed, there are four impact fuses that have a probability of $2E-4$ of at least one causing an inadvertent detonation. The specified probability of inadvertent detonation at mode change is $2.3E-5$. Determine the required probability of single-point failures to meet the specification.

REFERENCES AND BIBLIOGRAPHY

Bazovsky, I., *Reliability Theory and Practice*. Prentice-Hall, Inc., Englewood Cliffs, NJ, 1961.

Breipohl, A.M., *Probabilistic Systems Analysis*. Wiley, New York, 1970.

Raiffa, H., *Decision Analysis*. Addison-Wesley, Reading, MA, 1970.

Shooman, M.L., *Probabilistic Reliability: An Engineering Approach*. McGraw-Hill, New York, 1968.

Smith, C.O., *Introduction to Reliability in Design*, McGraw-Hill, New York, 1976.

IV

HAZARD ANALYSIS

21

INTRODUCTION

If we examine recent history, we will come upon shocking losses resulting from accidents. In 1986, there were 94,000 accidental deaths and 8 million disabling injuries in the United States. The motor vehicle was responsible for 47,900 deaths and 1 million injuries. The total cost of these accidents was at least $118 billion with $57.8 of that arising from the motor vehicle. Total civil aviation deaths in 1986 were about 1000. There were 10,700 deaths in the work environment with 1 million disabling injuries. Losses from the misuse or malfunction of major high-technology systems continues to be a major share of accidental loss in the United States. It is heartening to note that these figures are declining slightly in spite of increased exposure in almost every category. However, the magnitude of the loss continues to appall.

All losses from whatever source, particularly when they involve human misery, are to be lamented. Society should expend every effort to reduce unneeded losses from accidents. However, from the point of view of a government with limited resources, the larger losses will naturally command attention. It is particularly unsettling when the losses are due to the malfunction of technical systems that could have foreseeably been designed to function without such losses. The system safety professional is engaged in preventing these sorts of losses. Hazard analysis is the core of that work.

The earliest approach to safety was to examine the system during its operational life to correct what were deemed to be unacceptable hazards. This form of *a posteriori* safety, or safety after the fact, is deficient in two ways. The notion that an accident must be allowed to happen before it can be prevented is an archaic one, but unfortunately it is still abroad exacting an enormous cost in the operation of high technology systems. The other problem with this form of safety is determining what is acceptable and preventable.

During World War I, aircraft accidents were most often found to be unpreventable or to be merely the cost of operating such a dangerous system. Accidents caused by engine failure or structural failure were deemed unpreventable.

During World War II, we were aware that most aviation accidents were preventable but still found quite enormous losses acceptable. The loss of 8 fighter aircraft out of 24 during a three-month training period was considered lamentable but acceptable. And the pilots that survived such a training cycle were *ready*.

It was only with the advent of the ICBMs with nuclear warheads that an accident type was recognized that could not be allowed to happen even once. At this time, attention was turned to methods of hazard analysis that could detect potentials to do harm prior to the operation of a system.

The traditional or *a posteriori* form of safety has several other weaknesses that are corrected by the system safety form of hazard analysis. After the fact, leverage on the system is poor. The system is in operation, perhaps in great numbers. Changes will be costly. They must fit all of the operating environments of the system. Their installation will require withdrawing the system from service. Such high costs of safety changes will inevitably cause a lower state of safety to be acceptable.

Another weakness of this form of safety is that only part of the operational life of the system remains. If it is a large portion of the life, the safety change may be worth doing. If, as so often happens, an accident reveals a serious safety deficiency late in the system life, it may be deemed not cost effective to make the change. Accidents with their accompanying losses then continue to occur.

The foremost weakness in this form of safety is that of allowing a very severe accident to occure that will be catastrophic to the operators of the system and the industry it represents. Certain accidents have such large losses that the system operators should be willing to expend extremely large resources to prevent them; witness the cost of the Three Mile Island nuclear accident to the nuclear power industry.

System safety *imposes* a state of safety on the system throughout its life cycle. It is active during the design phase of system development where the leverage on the system is the greatest, where the cost of making changes to improve the state of safety is minimum. Discovering a hazard and analyzing it for cause during the system design phase must be followed by countermeasure implementation if the accident is to be controlled. The decision as to whether to make this safety improvement, even during the design phase of system development, should depend, in part, on the cost of the countermeasure compared to the present value of the losses saved over the life of the system. Other major factors in determining if a safety countermeasure has sufficient worth to warrant installation, are the *irreducibles* of the system and the accident process. Irreducibles are those aspects of an accident and its prevention that can not be quantifed. The loss of the good name of a company or organization as the result of a catastrophic accident is such an irreducible. These ideas will be discussed in depth in later chapters.

21.1 SYSTEM SAFETY ENGINEERING

It is possible to divide all system safety activities into two broad categories— *management* and *engineering*. Part I of this book describes the management activities.

At the risk of oversimplifying, the system safety engineering activities or tasks are those associated with the discovery and elimination or control of system hazards. The management activities or tasks are simply to assure that the engineering tasks are performed. A complete definition of system safety engineering may be set forth similar to that found in MIL–STD–882.

> System safety engineering is an engineering discipline requiring specialized knowledge to apply scientific and engineering principles, criteria, and techniques to identify and control hazards and their associated risk to an acceptable level.

Under this definition of system safety engineering, we may establish certain tasks that will be performed on high technology systems to assure the completion of the engineering function as follows:

Develop preliminary hazard list

Perform preliminary hazard analysis

Perform subsystem hazard analysis

Perform system hazard analysis

Perform operating and support hazard analysis

Perform occupational health hazard assessment

Perform software hazard analysis

Perform risk analysis on subsystems and systems and advise management on risk acceptance and control

System safety engineering should commence with system design requirements. The process continues through the design and development phases of the system life and to a lesser extent throughout all life cycle phases of the system. System safety engineering should be active throughout the design. Proper performance of the above tasks will necessitate a continuous process as many of these tasks are dynamic and must be modified and expanded as the system design proceeds and more knowledge is developed about the system.

System safety engineering is an included activity under system engineering and proceeds apace with it. Figure 21.1 illustrates. A partial list of system safety tasks to include management and engineering tasks is as follows:

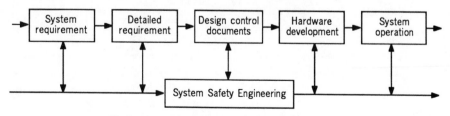

Figure 21.1   System safety engineering process.

1. Develop source documentation for system technology.
2. Create system safety standards, criteria, and requirements.
3. Familiarize with accident history of technology.
4. Prepare system safety program plan (SSPP).
5. Review engineering development with designers.
6. Attend design reviews.
7. Perform safety analysis on system as designed.
8. Review reliability data.
9. Develop recommendations for safety countermeasures.
10. Review test plans.
11. Prepare and document required and needed reports.
12. Review training plans and provide safety input.
13. Maintain safety data files.
14. Monitor safety record of operational system.
15. Provide system safety input to operational engineering changes.
16. Monitor system retirement for safety problems.

22

ELEMENTS OF HAZARD
ANALYSIS

As can be seen from the list of system safety engineering tasks in Chapter 21, hazard analysis comprises the major share of system safety engineering function. Hazard analysis contains two distinct areas, *hazard discovery* and *hazard control*.

A hazard may be defined as a potential for doing harm, while risk is *hazard evaluated*. The harm may take the form of injury, morbidity, or damage to property or the environment. Although an accident may be defined as an *undesired and unexpected event*, a hazard has another element attached to it that accounts for its uncertainty. The word *potential* accounts for the likelihood that a hazard will result in an accident. As system safety is, before the fact, safety, the system safety professional is not examining accidents but searching for potentials for harm, preferably in a system that is in the design phase of its life cycle. Having discovered these potentials, it is then necessary to analyze their causes in sufficient detail that a control strategy can be developed. Although the goal of the hazard control function is the elimination of the hazard, it is not often that a system can be so designed that it is impossible for an accident to occur. The control function takes the form of reduction in the magnitude of the hazard to an acceptable level. This control will manifest itself in only two fundamental ways—reduction of likelihood that a hazard will become an accident or reduction of the severity of the potential accident. The system safety professional must then possess the necessary skills and knowledge to discover hazards, analyze hazards for the causation elements, develop control countermeasures to reduce the hazards to acceptable levels, and to advise management on system risk levels that may be acceptable.

It has been the practice in some quarters to think of hazards as the potential for energy transfer in an undesired manner. Although this may be the case in a large number of hazards and in certain types of hazards this definition may be helpful, it is not true in all forms of hazards. It is not useful to rigidly adhere to this definition

197

of a hazard. The description of corrosion as an energy transfer is true in only the most subtle sense.

22.1 HAZARD SEVERITY

One measure of the degree of a hazard is its severity. The transformation of a potential to do harm into an accident will result in some degree of harm. Figure 22.1, depicts the elements of hazard severity or harm.

The magnitude of harm starts with a description of the immediate *consequences* of the accident. This should take the form of describing the type of event that has taken place. A "gear up" landing, a destruction of a booster that is straying too far from its launch trajectory, a release of radioactive water would be such a description.

The second step in describing the magnitude of accident severity is a description of the *losses*. A "gear up" landing of an aircraft can have a variety of results. In certain military aircraft carrying two external fuel tanks under the wings, there will be minimal losses. In the case of a passenger-carrying aircraft, the losses can be extensive, particularly if fire breaks out. Within the same type of aircraft, the magnitude of losses may vary considerably. The losses in a gear-up landing of a passenger carrying aircraft may vary from a few injuries incurred during evacuation to many deaths from fire. Losses can be imposed on the human as mortality, morbidity, or injury, on property, or on the environment.

The third and final step in describing the magnitude of severity is the *cost* of the accident. The costs may be somewhat variable, even given the same losses. The cost of injury may vary given the nature of compensation settled on the injured parties. The cost of destruction of a rocket booster may vary with the urgency of placing the payload in orbit.

We may see, then, that a complete description of severity of an accident is a set of three logically linked stochastic variables. However, the description of severity may be adequately described for particular circumstances by any of the three types of descriptors. It is also helpful to note that these three levels of description flow from the most simple and general to the most precise. In the more preliminary forms of safety analysis, the consequences may adequately describe the accident severity. In precise risk analysis, it will be necessary to estimate the costs of all hazards in the system.

A system safety program plan may specify that a system will be designed to assure no inadvertent launch of a missile. Although the system engineers may question the feasibility of making an inadvertent launch impossible, there is no

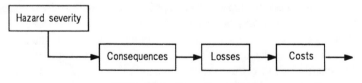

Figure 22.1 Hazard severity.

question as to the need to define the accident event severity more precisely. Still the magnitude of the event is quite clear from this statement. At the other end of this spectrum we may define a category A accident as one in which the total costs are more than $500,000. The Department of Defense document MIL–STD–882 defines categories of severity of hazards in four levels of loss, I, II, III, and IV, with I being the most severe, death of personnel or system loss, and IV being less than minor injury, morbidity, or minor system damage. Hazard severity is one of the two major measures of a hazard that enables the safety professional and upper management to judge risk acceptability. Of course, a judgment of risk acceptability cannot be made solely on the basis of severity. There is a finite, though very small likelihood of all four engines on a transport aircraft failing on a transoceanic flight. However, the likelihood of such an event is sufficiently small to make such a risk acceptable. There is also a very small likelihood of any nuclear power plant melting down and releasing its radio activity to the surrounding environment. For many, there is no likelihood that is sufficiently small to make such a risk acceptable.

An organization may develop its own categories of hazard severity so that they best suit the goals and functions of the organization. If one is doing risk analysis involving the evaluation of total life cycle system risks, discounting of future accident losses will be necessary and the costs of the hazards would be the proper measures of severity.

22.2 HAZARD LIKELIHOOD

Hazard likelihood refers to the probability of occurrence of an event that will cause harm. Such a likelihood may take the form of a hazard probability or of a likelihood category.

Early in the system development, hazard probabilities can not be developed due to a lack of information about system functions. At this time, broad hazard likelihood categories will be the only kind that can be established. Perhaps only three categories will be feasible at this time: *improbable*, *occasional*, and *frequent*. A fourth category of *impossible*, is suggested, but one must remember that an impossible hazard is not a hazard.

Later in the system development cycle, more definitive hazard categories may be used: *improbable*, *remote*, *occasional*, *probable*, and *frequent*. These are the categories of MIL–STD–882 and are perhaps the preferred categories. However, we must remember that when categories of likelihood are used, they must be further defined with words so that the analyst can judge their relative frequency and their effect on risk acceptability. For instance, an *occasional* hazard may defined as *likely to occur sometime in the life of an item*. Comparing this to a description of *remote* as *unlikely but possible to occur in the life of an item* clearly delineates between these two categories of likelihood.

One must always be aware that likelihood of an event occurrence cannot be defined by only the description of the likelihood of a single event. *Exposure* of the system or component to the possibility of a hazard must always be considered. The

Table 22.1. Hazard likelihood categories

Description	Category	Single Item or Event	Larger Exposure
Frequent	A	Will occur frequently	Continuous
Probable	B	Will occur several times in life	Frequent
Occasional	C	May occur once in life	Occur several times in life
Remote	D	Possible to occur in life	May occur once or more in life
Improbable	E	Assumed not to occur in life	Possible to occur in life

greater the exposure, the more probable the category. Thus, an individual may drive to work on a given day free of concern that his or her automobile engine will fail, leaving the individual stranded on the roadway for many hours. However, it a certainty that many engines will fail while the workers in a large city commute to work on a given day, and many more when all the commuting days of a year are considered. We may believe that a single component is reliable enough to place its failure likelihood category as *improbable*. However, if thousands of these components are in service in thousands of systems, each used thousands of hours per year, its category of likelihood may become *probable*.

Table 22.1 shows categories of likelihood with associated descriptions for two levels of exposure. These categories are quite similar to the categories of MIL–STD–882.

22.2.1   Hazard Probability

Referring to the statistical chapters earlier in this book, we can see that probability is a very precise measure of likelihood. Although we can apply the term "probability" to categories of likelihood, to be quite accurate, we should not do so. Probability is properly defined only if it is related to an appropriate exposure. The exposure used should be the best measure of the opportunity for the specified event to occur. For instance, we may consider the probability of an automobile accident as a probability per vehicle registered. It would, however, be more proper to speak of the probability of an automobile accident as a probability per mile driven. The miles on the road generate the opportunity for the accident rather than the simple ownership of a vehicle that may sit idle and safe in a garage.

Probabilities are most commonly calculated per unit time of use. But as in the case of the automobile, this is not always proper. It would most certainly not be proper to cite a probability of aircraft gear-up landing accidents as a probability per flight hour when it is the landing that generates the opportunity for this type of event.

Rates and probabilities are closely related and may be the same value when their magnitude is small. Of course, rates may exceed the value 1.00 as much as the circumstances dictate while probability is limited to a value of 1.00. Thus, at large values of an event rate, the probability of the event will approach the value 1.00 asymptotically.

It is difficult to know precisely the probability of any random event. However, simple experiments allow us to estimate a probability number to any degree of accuracy desired. Increased accuracy merely requires additional testing and increasingly rigorous analysis. The resources available for such testing then determines the accuracy with which the probability may be specified. Confidence in probabilities is a way of coping with their uncertainty. If an upper confidence bound on the probability of an accident event meets an acceptability standard, then the true value must also be acceptable.

Numerical probabilities must be used if in-depth risk analysis is to be performed on a system. Nevertheless, categories of likelihood are appealing even when good estimates of numerical probability are available.

22.3 HAZARD CONTROL

As we have previously discussed, a hazard possesses two properties—likelihood and severity. Together, these two properties measure the magnitude of a hazard and provide the proper measure of its control priority. A system may not tolerate a hazard of great severity even through the likelihood was very small. Of course a sufficiently small likelihood will make any hazard acceptable. The property of the Chernobyl accident that made it unacceptable was the degree of likelihood for this type of accident that a facility of this design engendered. The severity of the accident, given that it occurred, could not be mitigated without clearing the land for many miles about the facility. As in the case of the failure of all engines of a transport aircraft, this severity cannot be reasonably mitigated. It is the likelihood that determines the priority for control.

The foregoing is cited to caution against a tendency in some quarters to prioritize hazards solely on the basis of severity. Recall that we have stated that risk is a hazard evaluated. It is this evaluation that is the proper measure with which to establish the need to control a hazard to a more acceptable level.

If we are concerned with hazards that are the result of small electromechanical component failures, we may expect that the probability of all such failures will fall within a predictable range. This failure likelihood is normally such that accidents resulting from the failure of one such component are sufficiently likely that they should be prioritized by their severity.

If either of the two properties of a hazard has no value, the hazard is intrinsically safe or is simply not a hazard. If both values are other than zero, then the hazard must be examined for acceptability and perhaps controlled to a more acceptable level.

As system development proceeds, changes to improve the state of safety become more difficult and costly. Early recognition of the hazards and system modifications to reduce the risk of these hazards to an acceptable level is essential. As hazards are recognized, there is an order of precedence in the hazard control process that should be followed by system engineers and managers in reducing the risk of these hazards. The precedence in this hazard control process is the following:

Design for minimum risk
Incorporate safety devices
Incorporate warning devices
Employ procedures and training

Minimum risk implies a system design in which the risk of all hazards is minimum. This is a broad and powerful statement. Frequently there are insufficient resources to literally comply with this function or one could say that with increased resources the risk can always be reduced. However, the function means that there should be an attempt to fully eliminate all hazards through a risk free design. Failing to do this, all hazards should be controlled through the design process to assure system risk is at an acceptable level.

Acceptable risk is difficult to define. Risks that are acceptable in one environment are not acceptable in another. The risk of skiing is not acceptable in the home. The risks in a manufacturing process in the shop, are not acceptable in the office. We will discuss in detail the elements of risk acceptability in Part V. Briefly, risk acceptability depends on the environment in which the risk is incurred, the nature of the system and the necessity for its operation, the benefits of system operation, the perception of the nature of the risk, and the cost of risk mitigation.

Safety devices are the additions to the system principally for the purposes of improved safety that automatically control risk of hazards. If the identified hazards can not be eliminated or controlled to an acceptable level, safety devices should be employed to do so. These devices should take the form of safety design features that are a permanent part of the system and function automatically without human interaction. The devices may control the risk of the hazards by controlling the likelihood, the severity, or both.

A cable attached through a garage door-balancing spring to prevent the failed spring from exploding and causing damage to property or people is a safety device. It functions by controlling the severity of the hazard. A pressure relief valve is a classic safety device that again functions by reducing the severity of hazard. The valve functions to relieve the system pressure when system malfunction allows the pressure to exceed a safe maximum. The value does not influence the likelihood of the overpressure condition. A system that senses low altitude and airspeed and lowers an aircraft's retractable landing gear without pilot action is a safety device. This device functions to reduce the likelihood of the hazard.

Warning devices are the additions to the system principally for the purposes of improved safety that control the risk of hazards by warning a human operator and requiring further action by the operator to reduce the risk. If the device contains an automatic feedback loop that modifies the system for an out-of-tolerance operation, it should be considered a safety device. Warning devices are incorporated into the system when it is impractical to control system risks to an acceptable level with safety devices. Since warning devices require human response, and should be designed to minimize the possibility of human error in that response.

A light that comes on warning of overpressure in the system is a warning device. It will control the hazard if a human responds to the warning. A light warning of aircraft landing gear being in the up position under some established preset conditions is also a warning device. Warning devices may be audio or visual. Meter readings that become out-of-tolerance may be considered warning devices. However, in most cases, they should be supplemented with a more explicit warning of abnormal conditions.

Employ procedures and training to control the risk of hazards if all else fails. Procedures and training are the least effective of the four methods of control of risk in hazards. This technique must be used as a last resort recognizing that it will not be fully effective. An operator may be *trained* in the *procedure* of reading a temperature gauge at regular intervals and trained to take certain actions if the gauge is out of tolerance. Because human error rates will usually exceed that of the failure of electromechanical devices, this form of hazard control can not be considered satisfactory for hazards of high severity.

Procedures may be backed up by certification of the training or requiring a second person monitor the procedure of the primary actor. However, procedures lack the positive degree of control that the other methods employ. A procedure that suggests periodic inspection of the level of hydraulic fluid to prevent brake failure does not prevent a number of brake failures caused by low fluid. A low-level fluid warning light would be more effective. Designing brakes that did not require fluid or from which fluid could not escape would be the most effective control of this risk.

22.4 HAZARD INDEX

The degree of risk of a hazard is a measure of disutility or undesirability of the hazard. This disutility is a function of the likelihood and severity of the hazard. The precise way of measuring risk is with expected value. This is the product of the probability of an event and the value of the event. Failing to have precise probabilities and hazard severity values, a risk matrix may assist in hazard risk evaluation. We have noted in this section how certain hazards of high severity are acceptable because of very small likelihood. It is also true that hazards of low severity that occur with great frequency may be unacceptable. A matrix is an appropriate way of displaying the combinations of likelihood and severity of a hazard so that acceptability may be judged. Table 22.2 illustrates the principle.

In a matrix as that shown in Table 22.2, the high and unacceptable risks are in the upper left quadrant of the matrix. Risks that may be acceptable are in the lower right quadrant of the matrix. In any matrix, it is necessary for system managers to determine the risk levels that are initially unacceptable and thus must have one of the four methods of hazard control applied against them. The matrix facilitates the combination of levels of likelihood and severity as in the more precise measure,

Table 22.2. Hazard risk assessment

	Severity	
Likelihood	High	Low
High	High (high-high)	Moderate (high-low)
Low	Moderate (low-high)	Low (low-low)

expected value. MIL–STD–882 contains two examples of possible risk assessment matrices. Table 22.3 is another suggestion.

In this matrix, the four severity and five likelihood classes of MIL–STD–882 are assigned numerical values from the highest and most severe and probable to the lowest and least probable. Products can then be taken of combinations of likelihood and severity as is done when expected value is calculated. These products become a quite good relative assessment of risk categories in various system hazards. A value of 20 would undoubtedly be unacceptable and a value of 1, perhaps, acceptable.

The system safety professional can use such a matrix to examine the various system hazards in an organized fashion. The method is quite subjective in the assignment of likelihood and severity to their appropriate classes. Development of organizational standards will assist in this process. There are, of course, irreducibles that overlay such a process and effect decisions as to which risks are acceptable. Irreducibles usually pertain to moral values or human suffering. Familiarity with an organization's product line will facilitate a uniform evaluation of the irreducibles.

22.5 HAZARD CONTROL DESIGN CRITERIA

System safety design requirements are based on sound engineering practices and safety design principles for a particular type of system. There are general design criteria that may be followed on complex, high technology systems to reduce the degree of hazardous risk. A few are listed below.

Eliminate or control hazards by material selection. When potentially hazardous materials must be used, use those that serve the need while having the least risk.

Table 22.3. Hazard risk assessment matrix

	Severity			
Likelihood	4	3	2	1
5	20	15	10	5
4	16	12	8	4
3	12	9	6	3
2	8	6	4	2
1	4	3	2	1

Isolate hazardous substances, components, and operations from personnel and incompatible materials.

Locate equipment so that access during operations, servicing, maintenance, or repair minimizes personnel exposure to hazards such as electrical, toxic substances, radiation, and sharp injury-producing surfaces.

Attempt to minimize risk resulting from excessive environmental conditions such as temperature, pressure, noise, toxicity, acceleration, or vibration.

Design to minimize risk created by human error.

Consider alternate approaches to minimize risk from hazards that cannot be eliminated. Such approaches include interlocks, redundancy, fire suppression, protective clothing and equipment, devices, and procedures.

Protect power sources, controls, and critical components on redundant subsystems by physical separation or shielding.

Attempt to minimize the severity of personnel injury or damage to equipment in the event of a mishap.

Design software-controlled or -monitored functions to minimize hazards resulting from software malfunctions.

In sum, hazard analysis encompasses a set of methodologies that first searches for potentials to do harm in a system. Having found these hazards, further analysis attempts to control them to an acceptable level. To do so requires an understanding of the causes of these hazards. So hazard analysis attempts to determine the sets of primary events in the hazard generation process. Having discovered these events, it attempts to modify their likelihood or logical relationship in ways that reduce the risk of the hazard to an acceptable level. Hazard analysis also attempts to reduce the severity of accident events by protective devices and equipment, procedures, and forms of system modifications that reduce the magnitude of human and property damage in an accident event.

Some methods of hazard analysis are directed primarily to the discovery of hazards. Other methods find cause of hazards. Some do both. The types and methods of hazard analysis are described in succeeding chapters.

23

PRELIMINARY HAZARD ANALYSIS

Preliminary hazard analysis (PHA) is the initial effort in hazard analysis during the system design phase of the system life cycle. Its purposes are to identify safety critical areas within the system, identify and roughly evaluate hazards, and begin to consider safety design criteria. It is primarily an analysis of hazard discovery. It is a first and most important examination of the state of safety of the system.

This type of hazard analysis is performed early in the concept cycle of system development and performed repeatedly thereafter as the system is designed and defined. In military development the initial PHA may be performed by a group other than the contractor responsible for system development. This may be a military organization that will receive and use the developed system. However it is performed, it is very important that this analysis be done as it is management's first look at the risk that may be expected in the operation of the system. The results of the analysis will be useful in examining design alternatives in tradeoff studies.

Concurrent with or perhaps preliminary to the PHA, a preliminary hazard list (PHL) may be developed. The PHL is a listing of possible hazards inherent in a system of this type. Such hazards may arise in the performance of the PHA or independent of the PHA. However, hazards on the PHL will normally be analyzed by the PHA. The PHL, in its initial form, is developed from experience with other systems of this type. If it is to be updated as additional hazards are found in the course of system design, it should contain all of the hazards on the PHA.

The PHA should consist of at least the following activities:

1. A review of historical safety experience in similar systems.
2. An examination of basic energy sources. Provisions that will control these sources of energy must be carefully considered. These sources should include

fuels, propellants, lasers, explosives, other energy-containing chemicals, hazardous construction materials, and pressure systems.

3. Identification of the safety requirements and regulations pertaining to personnel safety, environmental hazards, and toxic substances with which the system must comply.

4. Safety-related interface considerations among subsystems and system elements should be examined. These considerations might pertain to material incompatibilities, electromagnetic interference, fire or explosion initiation, hardware and software controls.

5. Examination of environmental hazards such as shock, vibration, extreme temperatures, noise, exposure to toxic substances, electrostatic discharge, lightning, ionizing and nonionizing radiation to include laser radiation.

6. Examination of operating, test, maintenance, and emergency procedures for hazards that may arise in these activities. These would include human failures in operator functions, effect of factors such as equipment layout, lighting, exposures to toxic materials, effects of noise, vibration or temperature on human performance, life support requirements in manned systems, egress from manned systems during accidents as well as rescue and survival of human operators.

7. Examination of major facilities support equipment that will be necessary to operate the system. Such considerations as storage, assembly, checkout, or proof testing of hazardous components of the system must be examined. These will include toxic, flammable, explosive, corrosive or cryogenic fluids, radiation or noise emitters, and electrical power sources.

8. Examine software modules in their interfaces with hardware, operators, or other software for possible hazards. Special attention will be given to safety critical software commands.

9. Examine safety related equipment for adequacy. These will include interlocks, redundancy, fail-safe designs in hardware or software, fire suppression systems, personal protective equipment, ventilation, and noise or radiation barriers.

Figure 23.1 illustrates the life cycle phases in which the PHA should be performed.

The hazards set forth in a PHA are the product of the cognizant safety engineer's imagination, experience, and knowledge of the system and its operating environment. There are no logical processes or modeling methods that will develop these hazards or assure that they are all discovered.

As can be seen in Figure 23.1, the analysis is started early in the system life cycle when information about the system is incomplete. The safety analyst should, therefore:

1. Examine design sketches, drawings, and data describing the system and subsystem elements.

2. Consider functional flow diagrams and related data describing the sequence of activities, functions, and operations in the operation of the system.

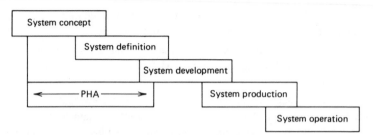

Figure 23.1   Preliminary hazard analysis in system life cycle.

3. Gather background information related to safety requirements associated with the contemplated testing, manufacturing, storage, repair, and use environments and safety related activities or problems of previous similar programs or activities.

23.1   FORMAT OF PRELIMINARY HAZARD ANALYSIS

It is convenient and usually standard practice to document the PHA on a columnar form. The exact nature of the form will vary with the preference of the safety engineer and system management. However, whatever the form adopted it should contain some minimum set of information.

Nomenclature of part or subsystem affected

System-operating mode for the threat or the hazard

A failure mode if component hardware is involved

A measure of the likelihood of the hazard

A description of the hazard

A description of the effects of the accident occurrence

A category of severity of the hazard

A general description of methods of hazard control

Figure 23.2 is a suggested example of a PHA form.

A brief description of each column entry will assist in understanding the form and its contents. Prior to column #1, it may be convenient to insert a reference column for the PHA file. This would be particularly useful if the PHA is stored in a computer file.

Column 1. The formal name of the part or subsystem in which the hazard originates or in which the hazard resides, should be entered. If no part or subsystem is involved, a procedure or system operation may be described in this column. Do not enter the part or subsystem that will be damaged by the hazard unless this element is also the origin of the hazard.

Program _____ System _____ Contract number _____

#1 Nomenclature or Part or Subsystem effected	#2 Operating Mode	#3 Failure Mode	#4 Estimated Probability	#5 Hazard Description	#6 Hazard Effects	#7 Severity Category	#8 Recommended Control	#9 Amplifying Remarks

Figure 23.2 Format of preliminary hazard analysis.

Column 2. The system operating mode during which the hazard is active is described here. A particular operating mode may apply to several of the elements of column 1. Each entry of column 1 may generate hazards in each of several operating modes. The operating mode will be that which pertains to the entire system when the hazard is active.

Column 3. Enter a brief description of the failure mode of the component or procedure that fosters the development of the accident event from the hazard. More than one failure mode may be cited for each entry in column 1 and for each operating mode of column 2. As a new operating mode or failure mode is developed, they each command a new line across the remainder of the form. Recall that a failure is not a hazard, but only a cause of the hazard.

Column 4. Enter the estimated likelihood that this hazard will manifest itself in an accident event. The likelihood may be a probability number such as $2.1E-3$ per hour. If a probability number is used, its exposure reference must always be given. It may be a category of likelihood described with a qualitative descriptor such as "improbable." If such descriptors are used, they must be defined in notes on the bottom of the form. The notes may define the categories in probability numbers or in words such as "unlikely to occur, but possible." The likelihood descriptors can be quite broad such as, "low, medium, high." Again, these categories of likelihood will require more precise descriptions. These likelihood descriptions are placed on the form to enable a manager not familiar with safety standards and practices to understand the nature of this hazard.

Column 5. The hazard description given in this column is the description of the events that lead to personal injury, morbidity, or property damage. The hazard description of this column is closely related to and normally differs from the failure mode of column 3. The hazard descriptions of several rows in the form may be the same. This is allowable if one of the preceding columns in each row is different (see example) (Table 23.1).

Column 6. Column describes the effects of the hazard on personnel and property. Care should be taken to avoid conflicting entries with column 5. Also, assure that column 6 is hazard effects whereas column 5 is hazard description.

Column 7. The severity of a hazard may be classified in categories or as a value of magnitude of loss. The categories may be the four general categories of MIL–STD–882 or somewhat finer-grained categories that might be preferred. Whatever the categories used they must be explained in notes on the form. Do not expect the management reader of this form to have 882 available.

Column 8. Columns 8 and 9 will usually pertain to control of the hazard. Column 8 may suggest some direct method of control. It need not be in detail, indeed, there is not space for a detailed description of a method of control.

Column 9. Column may be any comment that did not properly fall in the other eight columns. However, this column can be used to cite regulations, standards, or procedures that are directed to the control of this hazard.

Table 23.1. PHA of pressure system

#1	#2	#3	#4	#5	#6	#7	#8	#9
Tank	High pressure	Tank failure at pressure < design max.	Remote	Tank explosion	Injury to personnel, damage to equipment	I: nearby personnel and equipment; II: distant personnel and equipment	Isolation of tank	Analyze tank strength
		Tank failure due to overpressure	Occasional	Tank explosion	Injury to personnel, damage to equipment	I: nearby personnel and equipment; II: distant personnel	Isolation of tank; additional safety devices	Analyze for likelihood and acceptable hazard

211

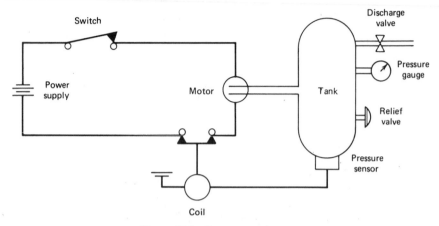

Figure 23.3   Pressure tank system.

It is obvious that the confined nature of the form encourages brief entries. Use words or brief phrases in documenting this analysis even though the analyst's notes developed in the conduct of the analysis are voluminous.

It may be that other columns suggest themselves as the analysis progresses. They can be added, however, one should remember that it is desirable to show all columns on a standard page.

23.2 PHA EXAMPLE

A PHA example performed on a simple system will illustrate the method. Figure 23.3 is a pressure system. Its function is to provide a constant source of high pressure on demand through the normal discharge valve. The switch is normally open. The operator closes the switch when it is noted that the pressure gage is below needed pressure. This provides power to the motor that contains a pump and will provide high pressure gas to the tank. If gas is not released and the tank becomes overpressured. The pressure sensor will sense the high pressure state and open the coil. This will interrupt power to the motor. Alternatively, the operator may sight the overpressure state on the gage and open the switch. If neither the operator nor the pressure sensor sense the overpressure state, the pressure relief valve will open relieving the pressure.

Two hazards of this system will be analyzed by the PHA shown in Table 23.1.

24

SUBSYSTEM HAZARD ANALYSIS

The subsystem hazard analysis (SSHA) is performed to identify design hazards in subsystems of a larger, major system. The analysis should find functional failures of the subsystem that will result in accidental loss. The analysis will examine component and equipment failures or faults, software failures or faults, and human errors that singly or in sets will establish a hazard due to the functioning of the subsystem.

A fault is a normal or inadvertent activation of a component or human function that singly, or in combination with other failures and faults, will result in accidental loss. The analysis shall include examination of all modes of failure of components and human operators that will propagate a fault chain through the system. The analysis techniques must be directed to finding both hazards and cause of hazards. The analysis methods may be either the inductive of the Fault Hazard Analysis (FHA) method, or deductive of the Fault Tree Analysis (FTA) method. Each of these analytical methods will be discussed in later chapters.

The analysis will pay particular attention to software within the subsystem that may generate a fault and resultant hazard. Care will be taken with software hazards to analyze and control them to an acceptable level.

The analyst will consult with design drawings, engineering schematics, and designers and related engineering personnel to perform this analysis.

This analysis or any analysis that discovers and reports hazards must have a risk threshold for hazards that are reported. Thus, some minor hazards may not be reported.

This SSHA should be first performed not later than the beginning of system definition phase of the system life cycle. As the system and related subsystems are further defined during system definition and development, the analysis should be revised until the beginning of the system production phase. Figure 24.1 illustrates.

213

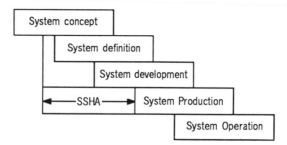

Figure 24.1 SSHA in the system life cycle.

24.1 SSHA EXAMPLE

A simple example will illustrate the SSHA method. Let us assume a command signal subsystem in a system that requires the command signal at an appropriate and precise time to continue to perform its function safely. Figure 24.2 is a schematic of the command signal subsystem.

Considering this subsystem, assume that there are two major malfunctions that will lead to a hazardous state: (1) inadvertent command signal or (2) no command signal.

Analyzing the subsystem with fault tree analysis for the events that will cause hazardous state (1), we find that there are 12 unique and nonredundant sets of events (cut sets) that will cause this state. The events may be labeled as follows:

A: Power on
B: Relay 1 fails closed
C: Relay 2 fails closed
D: Power switch normally closed by human operator
E: Command switch inadvertently closed by human operator
F: Power switch inadvertently closed by human operator
G: Command switch fails closed
H: Power switch fails closed

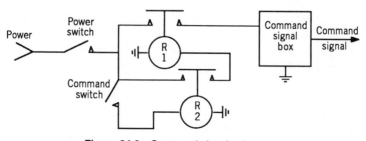

Figure 24.2 Command signal subsystem.

The sets that will cause this state are the following:

ABF, ACF, AEF, ABD, ACD, AED, AFG, ADG, ABH, ACH, AEH, AGH

Examining these sets we can deduce the following about this subsystem hazard. Event A appears in every set and is thus the most important in the causation of the hazardous state *inadvertent command signal*. However, it is a normal event and thus we must look elsewhere for control of this hazard. Events F, D, and H appear in four of the sets. Note these events all pertain to the power switch being closed. One of these is a normal event (D) and may not be dealt with to control this hazard. The other two (F,H) involve closing of the power switch by failure or by inadvertent human action. The remaining events—B, C, E, and G—appear in 3 of the sets. These are all abnormal events, one of which is a human failure.

In the absence of event rate information for these events, we may assume equal likely probability for all events. If this is done, we should focus our attention first on the power switch and events F and H. However, the inadvertent human action will undoubtedly be more probable than the failure of the power switch.

Next, turning attention to events B, C, E, and G, we see that event E is also a human failure and undoubtedly will have a higher likelihood than the other three events.

Next examining the system for the cause of the other hazardous state (2), a lack of a command signal, we can determine, again through fault tree analysis, that this hazard is caused by 8 single events or single point failures. The events are labeled as follows:

A': Power off

B': Relay 1 fails open

C': Relay 2 fails open

D': Power switch normally opened by human operator

E': Command switch inadvertently opened by human operator

F': Power switch inadvertently opened by human operator

G': Command switch fails open

H': Power switch fails open

The sets that will lead to hazardous condition (2) are the following:

A', B', C', D', E', F', G', H'

Note that these events are each the logical inverse of the corresponding events of hazard 1, although they are not complementary. The causation sets are single events.

There is only one normal event—event D'. We may examine all other events for control of this hazardous state. We note that events E' and F' are human failure events and may be expected to have a higher likelihood than the component related events.

A system in which we have 8 single events that can cause a hazard of any severity is usually in an acceptable state of safety. As two of these events are human events with their attendant high likelihoods, the system must necessarily be redesigned for the control of hazard 2.

We must also observe an important aspect of the relationship between hazards 1 and 2. The occurrence of any one of the single events that cause hazard 2 will prevent the occurrence of a number of the *sets* of events that cause hazard 1. For instance, event A′ will prevent all of the sets that cause hazard 1 occurring and thus completely prevent hazard 1. Of course, A′ will cause hazard 2. Event F′ will prevent the occurrence of 4 of the sets that will cause hazard 1. However, although hazard 1 may occur through the occurrence of the other causation sets, it has been precluded by the occurrence of F′ and hazard 2. We can then observe that hazard 2 is the dominant hazard. Our redesign efforts should focus on hazard 2.

Having completed the preliminary parts of the analysis, we are ready to prepare the report, documenting the hazards and their causes in terms of the events that together and singly will cause accidental loss due to this subsystem failing to perform its function.

Before we proceed to the documentation of this analysis, we should note that we did not perform quantitative analysis on this subsystem. If quantitative data are available for the events, such analysis is always preferable. Thus, we did not consider a time constraint in which these sets of events must occur for the accident to result from each hazard. Not having specified a time constraint, in effect made the time constraint infinite.

We also did not perform an FHA on this subsystem, nor did we make use of a Failure Modes and Effects (FMEA) analysis in analyzing this system. The latter method of analysis is normally performed by reliability persons for reliability purposes. However, it is most useful in subsystem and system hazard analysis in highlighting critical components. The FHA is also a component oriented, inductive analysis that can be used to highlight components that are safety critical. These methods will be discussed in later chapters.

Documentation of the results of this analysis may take any of a number of different formats. There are several essential elements of hazard that must be included.

Description of each hazard
    System mode
    Subsystem mode
    Hazard description
    Hazard effects
    Likelihood or relative likelihood of each hazard (hazards included from SSHA
        and O&SHA)
Causation events of each hazard
    Identify events precisely as to mode and constraints

Interactions between hazards

    Highlight complementary events

    Describe the exact nature of hazard interactions

If quantitative analysis is performed it should be reported in a special section that includes

    Event occurrence rates

    Event repair rates

    System time constraints

    Source of this data

    Results of quantitative analysis in probabalistic measures

Evaluation of safety of the system as affected by the subsystem

Recommendations as to control of system hazards caused by this subsystem

25

SYSTEM HAZARD
ANALYSIS

The system hazard analysis (SHA) examines the entire system for its state of safety. In so doing, it must integrate the essential outputs of the SSHA. It should identify safety problem areas of the total system design including safety critical human errors, and assess total system risk. Emphasis is placed on examining the interactions of the subsystems. The SHA should examine subsystem relationships for:

1. Compliance with safety criteria specified in subsystem requirements documents.
2. Sets of hazardous events, independent or dependent to including failures of safety devices and common cause conditions or events, that can result in system hazards.
3. Degradation of safety of the system from normal operation of a subsystem.
4. Contemplated design changes to system or subsystem that will affect safety or increase system risk.
5. Software system or subsystem control functions that may adversely affect system risk by software faults.
6. Human control functions that may adversely affect system risk through human faults.

The SHA will include all hazards discovered in the SSHA as well as descriptions of these hazards. Subsystems cause hazards only as they affect system function, thus all subsystem hazards contribute to system risk. Subsystems that were not analyzed by the SSHA or miscellaneous system functions or areas that are not included in a subsystem, must be analyzed separately and integrated into the entire system.

The integration of subsystems may be done directly in the case of a system that contains a few subsystems. If the subsystems are many or have multiple interactions within the system, a more formal network integration method, such as series–parallel, should be used. For instance, a computer subsystem may have multiple functional links within the system and create complexities where subsystem physical links are few and simple. Care should be taken in such cases to logically relate subsystem functions within the system.

The SHA requires determination of system risk. Thus, the analyst must determine system hazards, the severity of each hazard, and the likelihood of each hazard. Determination of likelihood requires an exposure period for which the risk is to be determined.

A complete SHA of a complex system may require the use of hazard analysis methods such as FHA, FTA, FMEA, and series-parallel network techniques.

The SHA is performed during system definition and during system development (see Figure 25.1).

The documentation of the SHA should contain the following information.

Descriptive information
  System mode
  Subsystem mode of subsystem of hazard origin
  Hazard description
  Hazard effects
  Likelihood or relative likelihood of each hazard
Causation events of each hazard
  Identify events precisely as to subsystem mode, system mode, and environmental
    constraints
Subsystem interface problems of special significance
  Identify subsystems involved
  Identity system and subsystem modes
System risk evaluation
  Severity listing of each hazard

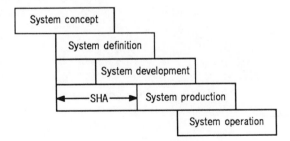

Figure 25.1   SHA in system life cycle.

Likelihood of each hazard

If full quantitative evaluation is conducted it should be presented for each hazard as in the SSHA

Risk summary, listing risks of each hazard and for the system as a function of system modes.

A logical evaluation of acceptability of system risk

Recommendations as to system risk control

The SHA may be performed more than once as the system develops, although it is not as dynamic as the PHA. The SHA is the cornerstone of the state of safety of the system.

26

OPERATING AND SUPPORT HAZARD ANALYSIS

The Operating and Support Hazard Analysis (O&SHA) is performed to identify hazards that may arise during operation of a system, to find causes of these hazards, to recommend risk reduction alternatives, and to impose an acceptable risk on the system. In the performance of the O&SHA, the following operating activities should be examined.

1. System testing, considering the scope of the tests, testing environments, and facilities or other systems employed during the tests.
2. Installation or modification of system, subsystem, or components during testing or normal operation.
3. Maintenance activities during system test or normal operations.
4. Transportation of the system either to its point of deployment or in the course of its normal use.
5. System storage either following deployment or prior to emplacement for normal use.
6. Accidents that may occur during operation or system test to include, emergency escape for personnel, fire control, environmental protection.
7. Postaccident problems including emergency medical care, toxic pollution, and system repair to include supporting facilities.

In the performance of the O&SHA, the analysis shall investigate certain special considerations that may influence safety of the system during its operational deployment.

1. Requirements for safety devices and equipment, including personnel safety and life support equipment. Examine the influence of these devices and

equipment on the risk level of the system considering the changes they may impose on the system.

2. Emergency procedures including warnings and cautions for egress and rescue shall be examined for adequacy.
3. Requirements for handling, storage, transportation, maintenance, and disposal of hazardous materials that are used or transported by the system should be examined for compliance with applicable regulations and an acceptable state of safety.
4. Requirements for special personnel training primarily applicable to safety should be examined. This would include necessary certifications and other regulatory requirements.

The documentation of the O&SHA shall contain the following information.

Descriptive information
    Operating mode
    Subsystems involved
    Hazard description
    Hazard effects
    Likelihood or relative likelihood of each hazard
Causation events of each hazard
    Identify events precisely as to subsystem, subsystem mode, system operating mode, and environmental constraints. Highlight human and software events.
System operating risk evaluation
    Severity listing of each operating hazard
    Likelihood of each operating hazard
    Risk summary, listing risks of each hazard and total operating risk
A logical evaluation of acceptability of system risk
Recommendations as to operating risk control

The hazards of the O&SHA will be included in the SHA, either as a central element of the documentation or an addendum.

27

FAULT HAZARD ANALYSIS

The Fault Hazard Analysis (FHA) is a method of analysis that may be used in the SSHA, SHA, or O&SHA. It is an inductive method that has the purpose of finding hazards that result from single-fault events in a system or subsystems that may arise from any system mode such as storage, transportation, or operation. It considers the effects of a single fault within a subsystem and across subsystem interfaces throughout the system. Because it traces the effects of a single-fault or component failure, it is sometimes called a single-thread analysis. The major and obvious weakness of the FHA is that it does not discover hazards caused by multiple faults. The FHA method is quite similar to the FMEA reliability analysis.

Since the FHA supports the SHA, SSHA, O&SHA, and, in some cases, a FTA, it is therefore performed at the time in system development when these analyses are performed.

The FHA is primarily an evaluation of the effects of subsystem component failures in all modes of system and subsystem operating modes and component failure modes. The effects of these failures are traced through the subsystem and subsystem interfaces to their final effect on the system.

27.1 FAULT HAZARD ANALYSIS FORMAT

The FHA documentation format reports the following on the system (see Table 27.1).

Component nomenclature
Fault originating condition
Component mode for fault origin
Subsystem mode
System mode

Table 27.1. Fault hazard analysis

Program ——————— System ——————— Contract Number

#1 Component Nomenclature	#2 Fault Condition	#3 Component Fault Mode	#4 Subsystem Mode	#5 System Mode	#6 Hazard Effects on Subsystem	#7 Hazard Effects on System or Mission	#8 Environmental Factors	#9 Secondary Factors	#10 Hazard Level[a]	#11 Hazard Control
Relief valve	Corrosive environment	Fails closed	High pressure	Operating	Overpressure	Tank failure	Corrosion	Proximity of persons	I: Remote	Inspect
		Fails open	High pressure	Operating	Lack of pressure	No pressure	Corrosion	Need for pressure	IV: Occasional	Positive seating of valve
Pressure sensor	Vibration	Senses high	High pressure	Operating	Low pressure	Insufficient pressure	Temperature	Need for pressure	IV: Improbable	Sensor test
		Senses low	High pressure	Operating	High pressure	Tank failure	Temperature	Reliability of pressure relief system	I: Improbable	Pressure and sensor relief test

[a]MIL–STD–882B severity categories. Probability categories: improbable, remote, occasional, likely.

Hazard effects on subsystem
Hazard effects on system or mission
Environmental factors
Secondary factors that may influence hazard effects
Hazard level
Hazard control

We will briefly describe each element of the FHA documentation.

Column 1. A simple, brief description of the component, normally a few words of nomenclature with part number if necessary for complete identification.

Column 2. A brief description of a fault originating condition if other than component originated.

Column 3. The mode of the component at fault origination, normally the failure mode.

Column 4. The subsystem mode that contains the component of fault origination.

Column 5. The system operational mode at the time of fault origination.

Column 6. A brief description of the effects of the hazard on the subsystem in which the fault originates.

Column 7. A brief description of the effects of the hazard on the system or the mission of the system.

Column 8. A description of environmental factors that may influence the likelihood or severity of the hazard.

Column 9. A description of other factors that may influence the likelihood or severity of the hazard.

Column 10. A brief description of hazard level, listing both likelihood and severity if set forth separately. An expected value of loss is a single-valued, complete description of the hazard level.

Column 11. Brief remarks as to a possible method of control of this hazard. The primary purpose of the FHA is to describe the hazard, not control it.

27.2 FAULT HAZARD ANALYSIS EXAMPLE

Table 27.1 is an example of the FHA as performed on the pressure system of Figure 23.3 for two components.

28

FAILURE MODE AND EFFECTS ANALYSIS

The Failure Mode and Effects Analysis (FMEA) is a reliability form of analysis usually performed by the reliability section. It may, on occasion, be structured to a safety outcome. The focus of the analysis is on single events that will cause a state of *unreliability* (or hazard) in the system. Such a reliability analysis may contain events that will not contribute to an accident. Nevertheless, the analysis is closely linked with safety analysis and system safety analysts must be aware of its function and methods.

The FMEA is the parent name of another and more important form of subanalysis called the Failure Mode Effects and Criticality Analysis (FMECA). Whereas the FMEA analyzes components of a system or subsystem for their contribution to a state of unreliability of the system, the FMECA attempts to find those components that are most important to this state of unreliability. These are called *critical* components.

Performing an FMEA on the pressure tank system of Figure 23.3, we might develop Table 28.1.

It is clear from Table 28.1 that the components together with their modes of failure will each result in different impacts on the reliability of the pressure system. For instance, the pressure gauge indicating higher-than-true pressure will result in low pressure but perhaps not so low as to cause the pressure provided by the system to be ineffective. On the other hand, the relief valve failing open will, most certainly, cause the system to be unreliable. Furthermore, it is seen that some failures have more impact on reliability than safety. For instance, the relief valve failing open is of critical importance for reliability but perhaps of no importance for safety.

The FMECA is an extension of the FMEA. Department of Defense MIL–STD–1629A, "Procedures for Performing a Failure Mode, Effects, and Criticality Analysis," describes the FMECA completely.

Table 28.1. FMEA of pressure tank system

Component	Failure Effect
Switch	Fails open—no pressurization
	Fails closed—possible overpressure if other components fail
Motor	Fails to function—no pressurization
Coil	Fails open—no pressurization
	Fails closed—possible overpressure if other components fail
Pressure sensor	Senses high—low pressure
	Senses low—possible overpressure if other components fail
Tank	Fails at normal pressure—system loss
	Fails at overpressure—system loss
Relief valve	Fails open—no pressurization
	Fails closed—possible overpressure if other components fail
Pressure gauge	Indicates high—low pressure
	Indicates low—possible overpressure if other components fail

Discharge valve: Not considered for reliability

A great problem in the performance of the FMECA and the FHA safety analysis is the computer control of system functions in modern complex systems. Such complex control paths defined only by microcircuitry and program statements are not readily analyzed by the cumbersome, manual, traditional methods. Automatic, computer-based methods are preferred and are available.

When extending the analysis to criticality, the criteria for criticality must be developed. These will be unique to each system and its use. However, they must be based on three elements:

Failure mode
Failure effect on reliability or severity of loss
Probabilities of the above

An FMECA worksheet should contain the following columns.

Component identification
Component function
Failure modes
Causes of failure modes
System and subsystem mode
Description of failure effects
    Subsystem
    System
    Environment
Failure level (mode, effects, probabilities)

Failure controls

Applicable standards and regulations

Having completed the worksheet, the analyst should then perform the criticality analysis. This analysis should result in a criticality coefficient for each component. Such a coefficient will be the expected value of severity or the expected value of loss if the analysis is being performed as a safety analysis, from the failure of each component in each of its modes. This coefficient will consider the probability that a component will fail in a given period of time (k), the conditional probability that the failure will be in a given mode (b), and the product of the severity or unreliable level and conditional probability of that level (c). The sum of the criticality coefficients C_s, will be the criticality coefficient for that component, C_c.

Probability values must be estimated for most of the calculations, but in performing a structured form of criticality analysis, the relative levels of criticality among components will be revealed. It can be seen that the analysis must take quite a different point of view if the outcome is to be safety rather than reliability.

In performing a reliability FMECA on the pressure tank example, it would be expected that the motor would be critical. This same analysis performed as a safety analysis would find the motor not-at-all critical. It would undoubtedly find that the relief valve, pressure sensor, or coil are most critical.

29

FAULT TREE ANALYSIS

Fault Tree Analysis (FTA) has, since its development in 1961, gained widespread recognition as one of the more powerful analytic tools for analyzing sets of events arranged in systems. These systems may lead to either desired or undesired outcomes. The method was first conceived by H. A. Watson of the Bell Laboratories to evaluate the safety of an ICBM launch control system. The method has been further refined over the intervening years so that it can evaluate ever more complex systems. There are available today computer programs that will perform the analysis following a synthesis of the Boolean logic structure of the tree by a human mind. Although the method is entitled *fault* tree it may be used to address systems that lead to desired outcomes. These may be called positive trees. D. F. Haasl was one of the early practioners and developers of the method. Others have made contributions (Fussell, 1977, Lambert 1975), and others.

The method structures relations between events in a system into a Boolean logic model that leads to accident causation. These events are structured so that they lead to a specified outcome. This approach to analysis is called deductive. The outcome is called a *Top event*. The method has the flexibility and power to perform this form of analysis when the more traditional form of series–parallel analysis breaks down. The method is unusually versatile in that it allows dynamic considerations to be considered, sensitivity analysis performed, and the results of analysis quantified. It allows the analyst to evaluate alternatives in system design—to judge tradeoffs. It is able to assimilate failure rates, down times, repair times, and other dynamic measures of system function.

The fault tree method has four major advantages over other forms of system analysis.

1. *It Directs the Analyst Deductively to Accident-related Events.* We have previously discussed inductive analytical methods such as the FHA. In this form of analysis, the component failure was assumed and the effect on the system

or subsystem determined by analysis. Thus the undesired system event occurred as a by-product of the component failure. The deductive approach assumes the major system or subsystem failure and examines lower order events to determine what are all combinations that could cause this specified, harm-causing event.

It is undoubtedly more useful to know all possible combinations of low-order events that will cause a major accident event than it is to know the impact on the system of a single component failing.

2. *It Provides a Depiction of System Functions That Lead to Undesired Outcomes.* A graphical depiction of the relationships between events that lead to a major accident is a comforting item for management to view. It is easy to read and understand on small to modest size trees. If the system is large and complex, the graphical depiction of the accident causation process may provide little information. However, it continues to be desired by management.

3. *It Provides Options for Both Qualitative and Quantitative Analysis.* Quantitative analysis is usually desired if quantitative data is available for the tree End events. Frequently, however, safety data are either not available or unreliable. In such cases, qualitative analysis will provide the accident causation event sets and measures of the importance of the individual End events in the causation process. In addition, quantitative analysis requires data on only the End events. No data need be provided on the intermediate events. Qualitative analysis results in sets of events that cause the Top event and a ranking of these events for their importance in causing the Top event. Qualitative analysis is more commonly used because it does not require precise event rates for the End events. It is quite useful in examining system logic for its causation of accident events. Quantitative analysis requires quantitative input data that are not easily obtained for systems under development. However, such analysis allows one to measure the likelihood of occurrence of the Top event or of any node or subsystem within the tree. This method also allows the analyst to determine a probabalistic measure of Importance of each End event in the causation process. Quantitative analysis is necessary if an objective measure of risk of system operation is to be obtained.

4. *It Provides the Analyst with Insight into System Behavior.* The process of fault tree analysis is so carefully prescribed and so detailed in its logical relationships, that it forces the analyst to understand the system beyond the level enjoyed by the system designers involved in only one part of the system or even system managers.

29.1 FAULT TREE SYMBOLOGY

The Boolean logic structure of a fault tree is depicted graphically by the tree diagram. The analyst proceeds from event to event asking at each level what is the cause of this particular event then showing the event types and their logical relationships in the tree diagram. Symbols are necessary both to depict the type of event and the nature of their logical relationships.

29.1.1 Event Symbols

Fault tree event symbols are of four fundamental types. Event symbols depict a fault event, a basic event, an undeveloped event, and a normal event.

A *fault event* is shown by a *rectangle*. It is always an intermediate event in the tree, never an *End* event or the last event in a tree branch. A fault event must be further described in terms of events that cause it. If, in the course of synthesizing the tree, one wishes to terminate a branch that has as its last event a rectangle or fault event, the rectangle must be changed to one of the End event symbols. If one decides that analysis is to be quantitative, a basic event will be assigned a rate of occurrence and, if repairable, a mean down time or repair rate. Rectangles are never assigned rates because they are never End events.

Fault Event

A basic event is shown by a *circle*. This is also an End event because there will be no further development of a basic event. The circle is frequently used to show the internally caused failure of a component. The analyst may choose to terminate a branch with a circle event though it is apparent that there are possibilities for internal analysis of the component for the causes of its failure. If the analyst chooses to further analyze a circle or basic event, he or she must replace the circle with a rectangle and continue the analysis. If the tree is to be quantified, a circle will be assigned an event rate and a mean down time.

Basic Event

An *undeveloped event* is shown by a *diamond*. An undeveloped event is an event that by its complexity invites further analysis, but the analyst chooses to terminate with a diamond. A subsystem may be shown as failed, or functioning in such a way as to transmit a fault, with a diamond. Human failures due to internal causes are shown with a diamond. For instance, a switch that a human operator is supposed to open at critical times in the operation of the system, may be shown closed in a diamond. There may be curiosity as to why the operator failed to open the switch. However, the analyst, sensing the difficulties of such an analysis, may choose to terminate that branch with a diamond. In quantitative analysis, diamonds are assigned rates and mean down times if repairable.

Undeveloped Event

A *normal event* is shown by a *house*. A house is sometimes called a switch event because it may be considered as occurring or not. The probabilities of a house event considered as a switch event would be 1.0 or 0.0. Typically, a house is assigned to power events. The quantitative analysis can then be conducted with the houses or switches on or off. Either states of events could be considered normal. A house can also be considered as either less than a certainty or greater than impossible. If the exposure time of the system for which the analysis is being conducted is long, the mean on or mean off state of the normal events may be used for quantitative analysis. Qualitative analysis can be performed with house events shown in the cut sets in either on or off state by events or their complements.

Normal Event

An infrequently used symbol is the *double diamond*. A double diamond indicates that the branch has been terminated with complexity not analyzed, but the analyst has the knowledge to continue the analysis. It is used when contributing events of tree branch are known, but the analyst believes it is not useful to show further analysis. As an End event, the double diamond must have a rate and perhaps a repair time for quantitative analysis.

Double Diamond

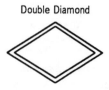

A *modifier* to an inhibit gate is shown in an *ellipse*. Such a modifier is always a condition that must be present for the inhibit gate to pass the fault. Conditions may be thought as states that occur over a *period* of time while events occur at a *point* in time.

Condition Modifier

A *triangle* is not in itself an event, but it represents an event. A triangle is a *transfer* symbol for an event. If in the process of synthesizing a tree an event in the working branch is found to be present in another branch, the analyst may transfer it to the other identical branch, terminating the working branch. A transfer symbol may also be used to continue the tree on another page.

Transfer Symbol

29.1.2 Logic Symbols

A fault tree represents logical relationships between events of a system. These relationships may be described by symbols representing the many possible logic forms the relationships may take. There are two basic logical relations with several subsets for each. These relationships are depicted as gates with shapes representing Boolean logic.

The logical OR gate is a logical relationship between gate inputs that requires the occurrence of *one or more* inputs within the time or exposure constraint for the gate to pass the fault. As depicted in Figure 29.1, one or all of the events A, B, or C will cause the output E. The distinctive shape of the OR gate conveys this meaning and must be maintained in the drawing of the tree. Quite strictly, the OR gate represents the Boolean union logic.

In some systems, if one event occurs the others are excluded from occurring. This form of a logic gate is called an *exclusive* OR gate. In calculating probabilities through OR gates, if we recall Boolean union we can understand the difference

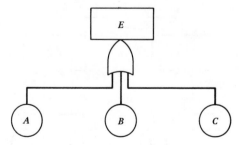

Figure 29.1 OR gate.

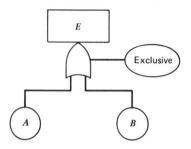

Figure 29.2 Exclusive OR gate.

between the two types of OR gates. A simple addition through the exclusive OR will give an exact probability of fault passage. In the simple OR gate, an addition will always give a probability equal to or greater than the true value. In the case of faults, such a probability will always be conservative. Figure 29.2 depicts an exclusive OR gate. Imagine event A of this gate as a relay failing open. If the fault will also be passed if the relay fails closed, then event B can represent that mode of failure. Event E will then simply represent relay failure. The simple modifier, exclusive, is sufficient if the events through the gate are mutually exclusive. An exclusive OR gate can be created with a set of simple OR and AND gates. The AND gates must show the exclusivity of the events. Figure 29.3 depicts the exclusive OR transformation.

A more complex form of OR gate is a *summation* gate. Such a gate is used to relate events that require levels of events to pass a fault. In the example shown in Figure 29.4, levels of pressure, temperature, and acidity in a petrochemical facility will pass the fault. Figure 29.4 illustrates the basic form of such a gate.

The summation gate depicted is one in which certain levels of pressure, temperature, and acidity are necessary to pass the fault. For instance, if a single parameter is high enough the fault will be passed. If pairs of parameters are somewhat less

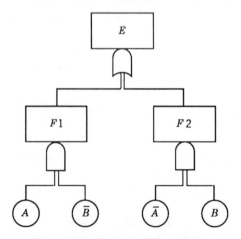

Figure 29.3 Exclusive OR resolution.

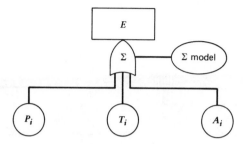

Figure 29.4   OR summation gate.

elevated but still well above normal levels, the pair will pass the fault. In a like manner, the three parameters at a somewhat lower level will also pass the fault. The following relations are the summation model.

$$T_i \geq \rightarrow E$$
$$P_i \geq \rightarrow E$$
$$A_i \geq \rightarrow E$$

$$\left.\begin{array}{l} P_1 \leq P_i < P \\ A_1 \leq A_i < A \end{array}\right] \rightarrow E$$

$$\left.\begin{array}{l} T_1 \leq T_i < T \\ P_1 \leq P_i < P \end{array}\right] \rightarrow E$$

$$\left.\begin{array}{l} T_1 \leq T_i < T \\ A_1 \leq A_i < A \end{array}\right] \rightarrow E$$

$$\left.\begin{array}{l} P_2 \leq P_i < P_1 \\ T_2 \leq T_i < T_1 \\ A_2 \leq A_i < A_1 \end{array}\right] \rightarrow E$$

Having specified the conditions for the summation gate, such a gate could be transformed by a computer or manually to a set of simple AND and OR gates. For instance the summation model as shown above would transform to the tree of Figure 29.5.

An *m-out-of-n* voting gate is another form of OR gate (Figure 29.6). In this form of gate it is necessary to have m or more events occur out of the total number possible events that may contribute to fault output. For instance, consider a navigation system that consisted of three inertial platforms and the loss of two would cause critical loss of accuracy.

This gate can be transformed into three AND gates through the OR gate, each AND containing two of the events, $A_{1...3}$.

The other principal logic gate symbol is the AND gate (Figure 29.7). This gate type indicates a logical relation that requires all of the input events to pass the fault through the gate. This gate is the Boolean logic intersection.

An example of the AND gate would be the case of a navigation system that had three inertial platforms. Loss of navigation capability would be the loss of all three platforms. Figure 29.8 illustrates such a gate where A_1, A_2, and A_3 are each the loss of a platform.

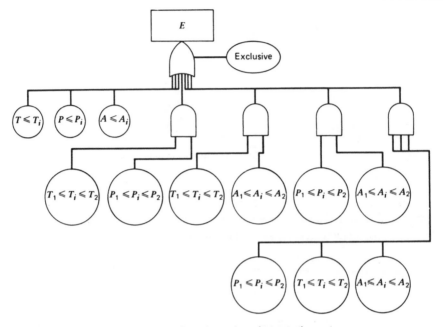

Figure 29.5 Transformation of summation gate.

A priority AND gate is an AND gate in which events must occur in a specified sequence. For instance, if a satellite system will power down if contact C is broken prior to opening relays R_1 and R_2, a *priority* AND gate will be the proper gate to depict this logic. Figure 29.9 illustrates this fault flow.

There are many other forms of OR and AND gates that can be used to describe forms of logic necessary for fault flow. However, the majority of these special gates can be simplified to simple OR and AND gates.

The *inhibit gate* allows the flow of fault if a condition is present and the input event occurs. Recalling that a condition is in place over a period of time as opposed to an event that may occur at a point in time, we may think of the condition as an enabling event. For instance, we may consider the case of the O ring seal that will fail if the temperature is below a specified value when pressure is applied. We could diagram this fault flow as shown in Figure 29.10.

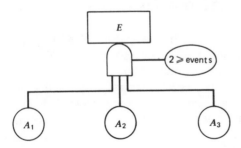

Figure 29.6 Loss of navigation accuracy.

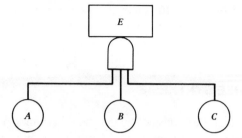

Figure 29.7   AND gate.

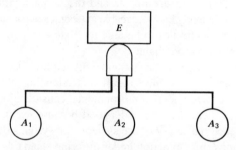

Figure 29.8   Loss of navigation capability.

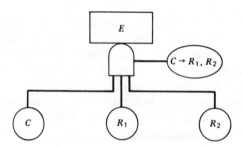

Figure 29.9   Priority AND gate.

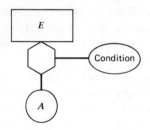

Figure 29.10   Inhibit gate.

The condition in Figure 29.10 is temperature less than the specified value. We can see that the temperature condition enables the output fault given the input initiator of high pressure.

29.2 FAULT TREE SYNTHESIS

The synthesis of a fault tree proceeds by the analyst asking repeatedly the cause of an event that appears on the tree. The response is carefully structured to include the classes or kinds of events for which the analyst is searching. The analyst must either have a thorough understanding of the particular system being analyzed or elicit this understanding from documents and discussions with system designers. The analyst must be sufficiently knowledgeable in technical areas such as aerodynamics, structural analysis, or electronics to be able to relate events that originate in these areas to the accident causation chain.

Such intimate knowledge of the system will require discussions with system designers and managers to develop a full understanding of the system and its operating modes and environments. Environmental considerations of the operational employment of the system must be understood because events generated in those domains may influence the safety of the system. Subsystems designed by different teams or organizations, may function under differing standards. Much of the value of fault tree development may come from understanding the system as the analyst probes the various design teams between which there may have been little prior communication.

The designer in attempting to simplify design solutions, may separate the design into subdesign problems. Such an approach may leave interactive gaps in system function understanding that the fault tree analyst must bridge. The fault tree strict logic structure and the analyst's viewpoint will uncover many of these design oversights prior to operational failure of the system.

Although computer codes are used extensively in analyzing the tree after it is synthesized, there is as yet no program that will snythesize a tree containing human or environmental events. There are codes that will synthesize systems containing purely hardware elements. However, the lack of intercourse between designers and analyst and the inability to consider human and environmental events, make such an approach undesirable and perhaps more of an academic exercise. There is a process called *directed graph* (digraph), developed by Lapp and Powers (1977) for chemical facilities and extended by Lambert, and later applied to transportation system by Roland and Philipson (1980). A tree may be computer synthesized from a digraph, however the digraph must be developed by human reasoning and the process is quite similar to synthesizing a fault tree.

The synthesis of the tree commences with the specification of the Top event. The Top event must be carefully specified considering the environment in which the system is functioning its desired mode of function and the system state in this environment. The Top event, inadvertent missile motor ignition would have a different tree if it was to be considered during storage rather than in place for launch.

A guidance subsystem failure would have a different tree if the system was being operated in the flight mode rather than the preparation for launch mode.

In selecting the Top event, careful care should be given to the system boundaries and the occurrence of conditions and events across these boundaries. If the Top event is to be turbine engine failure, the analyst must consider the state of the flows across the engine boundary. Such conditions as fuel flow and electrical signals to the engine must be assumed prior to synthesizing the tree. The normal approach to activity across system boundaries is to assume normal conditions and events.

Although a fault tree can contain normal events, the majority of events that appear in the tree will be either faults or failures. In discussing the events in the tree it would be proper to address them all as faults. Even normal events are faults if they are necessary to propagate the fault flow. A failure, like a normal event, can be considered a fault. However, there are more specific definitions of faults and failures that will assist in defining the events of the tree. There are five classes of these events that will be logically linked with the tree logic structure.

1. *Primary Failures.* These are component-related failures caused by problems internal to the component. Repairing a primary failure will return the component to a functioning state.

2. *Secondary Failures.* These are component-related failures caused external to the component. Environmental stresses such as temperature or vibration cause secondary failures. Repairing a secondary failure does not return a component to a functioning state.

3. *Primary Faults.* These are event occurrences that create fault flow that are not component related. They may be normal or human events. A primary fault may self-repair. A human failure caused internally will self-repair.

4. *Secondary Faults.* These are events that propagate the fault flow that are externally caused or generated. A human failure caused by external stress such as high ambient noise would be a secondary fault. A secondary fault may self repair, but the analyst should be cautious in so assuming.

   A malfunctioning component may be a failure or fault. A relay that fails open is a failure. A relay that opens inadvertently due to a command is a fault. A warning light that does not illuminate is a failure. A warning light that is not heeded is a fault.

5. *Command Fault.* A command fault may be defined as a fault or failure that is caused by commands external to the source of the fault. For instance, a command fault of a relay would be "relay activates inadvertently" or at the wrong time for safe system function. Such action that is commanded by other parts of the system would be a command fault. A human can be commanded to fault by being given improper information for action. Improper meter readings that trigger improper action on the part of an operator, would be a command fault. Repair of the direct cause of the fault, the improper operating component or operator will not repair the system. In this sense a command fault is like a secondary failure or fault.

The synthesis of a fault tree must be done according to a few well-tested rules. If the analyst proceeds directly and optimistically to put together events that he or she believes are causally related he or she will undoubtedly make errors in the logic as well as miss important events.

The synthesis of a tree starts with the *Top event* that has been determined to be one of the most undesired safety-related events that can occur in the operation of the system. The synthesis of a tree is a costly and time-consuming process. Therefore, trees for trivial events are not worthy of the effort. The Top event should be carefully considered as to the system state or mode of operation as well as the system boundaries. Furthermore, the analyst should consider the flows across the system boundaries. For instance, as previously mentioned, in the case of an engine fault tree the flows across system boundaries must be specified and would usually be normal for the state of engine. An engine at idle on the ground would have different boundary conditions than an engine operating at high power and at high altitude. The fault tree for the inadvertent detonation of a munition would be quite different if the munition were in a state of storage rather being carried toward a target.

A large system may be broken into categories of events immediately under a top OR gate. For instance an aircraft landing accident may be developed by dividing the tree into the types of landing accidents passing through a top OR gate. Such a tree will, in fact, consist of several fault trees, one for each of the landing accident types. The Top events of these subsystems will usually include subsystem faults or failures which in themselves will be substantial trees. In a like manner, a tree can be synthesized for a number of subsystem failures by having the Top event be the failure of any one of these subsystems.

Having specified the system and Top event the analyst proceeds to define the next level within the tree. This level will consist of the set of events that are the most *immediate necessary and sufficient* events to cause the Top event. It is important to look for the most immediate events and to do so by considering generic causes of the event that is being analyzed for cause. For instance, relay 2 failing to open could be a cause of failure of relay 1 failing to close. However, the analyst should merely consider "no power to relay 1" as the generic cause of the event. It is then possible to analyze for no power to relay 1.

The analysis commences with the Top event being examined for cause and works down through successive levels of the tree asking at each event: What are the most *immediate necessary and sufficient* causes of this event? However, prior to delving directly into causes, the analyst should ask another question of each event. "Is this a component-related event?" If the answer to this question is "yes," the analyst must place an OR gate and through the OR gate pass primary, secondary, and command causes of the event being analyzed. If the answer is "no, this is not a component-related event, but a system-state event," then the analyst is free to place any type of gate and type or number of events through the selected gate. The question of component-related event must be answered for the Top event and all other events in the system that are not End events. Figure 29.11 illustrates the resolution of a component-related event.

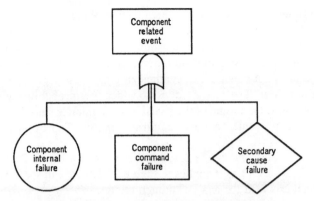

Figure 29.11 Analysis of a component-related event.

For instance, if the event to be analyzed is "Pressure Tank Fails," the analyst should define this event as component related. Thus, the event should have an OR gate with primary, secondary, and command events leading to that gate. The primary failure would be "Tank fails at Normal Pressure." The secondary failure would be failure of the tank due to some external cause such as projectile striking the tank or perhaps corrosion.

Figure 29.12 is a small system that we will fault tree.

This system functions by an operator closing switch. This provides power to relay 2 that closes relay 2. Movement of the relay contacts under power is shown by the arrow. Closing of relay 2 provides power to relay 1, which closes. This provides power through relay 1 contacts to the signal output box that gives command signal when under power.

We will assume the Top event "Inadvertent Command Signal" and synthesize the tree for this event. The system state is standby, as shown in the figure. The Top event is considered as to whether it is a component-related event. The answer is no. There may be some question in this case. The analyst can ask of himself or

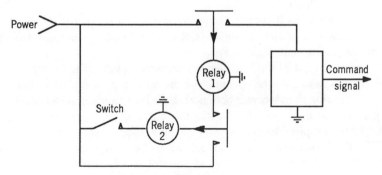

Figure 29.12 Electrical subsystem.

herself a related question to assist in this determination. *Will failure of the component cause the event*? In this case, the answer is "no." The failure will prevent the event.

This brings us to an important and, sometimes, troubling consideration in synthesizing fault trees; that is the "no miracle" rule. The no miracle rule states that the analyst does not consider, in synthesizing a tree, low probability events that prevent fault flow. The analyst does consider low probability events that cause fault flow. Thus, the failure of the signal box is a miracle since its occurrence will prevent the Top event. This also confirms that this Top event is a system state event.

Following these rules, the tree is developed as shown in Figure 29.13.

Examining Figure 29.13 as it applies to Figure 29.12, we can see how the rules of fault tree synthesis lead the analyst through the tree.

The Top event is *not* a component-related event and thus any type of gate is proper to this event. Examining the schematic of Figure 29.12, it can be seen that this signal can be caused by relay 1 being closed and power being on. These are the most *immediate* causes. At this point in the tree we are not concerned with relay 2 or the switch. Power on is a normal event and is shown in a house as an End event.

"Relay 1 closed" is a component-related event and thus must have an OR gate with primary, secondary, and command events to that gate. The secondary event is an environmental event. If one is not known, an S in a diamond suggests that there may be a secondary cause of this component related failure. The primary event is the internal failure cause of the event to be analyzed. In this case, "Relay 1 fails closed," is sufficient to describe the basic event internal failure that is a cause of "Relay 1 closed." The command event is shown as simply the command of relay 1 closed. The analyst may wish to leap ahead to the cause of the command event, however he or she is cautioned against this course. It is always best to take small, ministeps to avoid overlooking events.

Examining the schematic it can be seen that power and "Relay 2 closed" are both needed to command relay 1 closed. Note that the power event is repeated even though it appears in the tree under the top AND gate. It is necessary to close relay 1, so it should be shown. The resolution of the tree to its minimal cut sets with Boolean logic, will eliminate redundant events.

"Relay 2 closed" is a component-related event, thus primary, secondary, and command events are shown through an OR gate. Relay 2 is commanded closed by the "Switch closed" and, again, "Power on."

The switch closed event is a component-related event. The switch command closed event is caused by the operator. We are now at the boundary of the system. There is a question as to whether both normal and operator command events would be appropriate at this point. Recall that the Top event is "*Inadvertent* command signal." To be complete the "inadvertent" notation should have been carried through the tree in all the command events. Thus the switch closed event should be "inadvertent" which will cause the operator command event to also be inadvertent. The normal closing of the switch would cause a normal command signal. This is not the Top event.

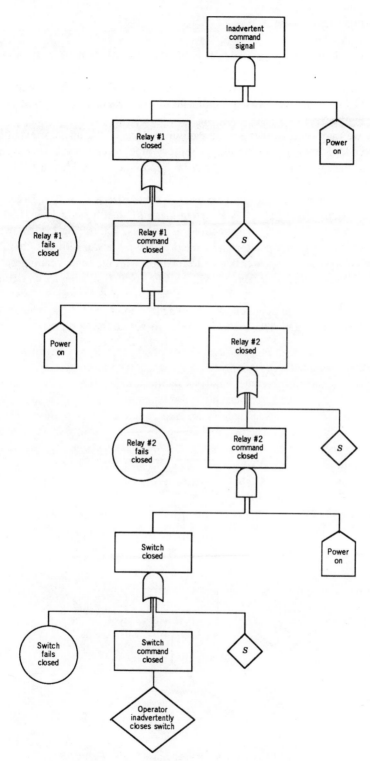

Figure 29.13 Fault tree of inadvertent command signal.

When analyzing operator events it is helpful to think of them as components. This means that the operator events will have as input an OR gate and primary, secondary, and command inputs at least considered. Examining the fault tree of Figure 29.13 again, the "Switch command closed" event can analyzed again as seen in Figure 29.14.

Now we may observe that a more careful analysis of the operator event brings forth the additional event of "Improper sensory input to operator." This event was missed in the previous version and may be important in considering the cause of the Top event.

These fault tree principles will be used to synthesize a tree for a larger, more complex system. Figure 29.15 is a system that periodically activates a motor.

The system of Figure 29.15 functions as follows. The spring-loaded switch is closed manually by the operator and is held closed by power in upper circuit. The timer relay on receiving power begins to function with the timer running out at 90 seconds. Power in the upper circuit will activate relay 1, closing relay 1 contacts. This event starts the motor. At 90 seconds the timer will activate and open timer relay contacts. This will open relay 1, stopping the motor. It will also release spring-loaded switch to open. The timer will reset to zero. The system is now set for another cycle.

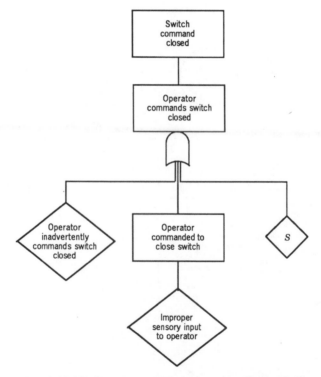

Figure 29.14 Operator command events (see Figure 29.13).

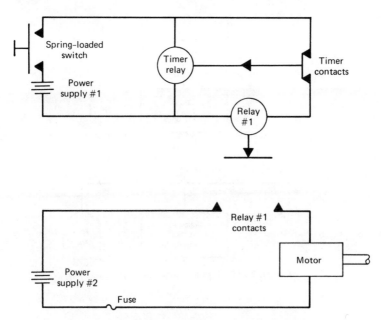

Figure 29.15   Schematic of motor control system.

The Top event of this system will be specified as "Motor operates $t > 90$ seconds." At first consideration, this might be thought to be a component related event. However, the failure of the motor would be a *miracle* and will *prevent* the fault flow and the Top event. The Top event is thus a system-state event that can be caused by "EMF to motor $t > 90$ sec." The "EMF to motor $t > 90$ sec." requires three events: "power supply 2 on," "Relay 1 contacts closed $t > 90$ sec.," and "Fuse closed." Note that the power supply and fuse events are normal events. Here again, the "no miracle" rule does not allow these normal events to fail. Figure 29.16 shows these first levels of this tree.

Analyzing the "Relay 1 contacts closed $t > 90$ sec." event, we note that it is a component-related event. Thus, we must place an OR gate with primary, secondary and command events through that gate. The "Relay 1 contacts command closed $t > 90$ sec." is further defined by a structuring event "EMF to Relay 1 $t > 90$ sec." This in turn is a system-state event. Examining the system schematic, we can see that it will be caused by "Spring-loaded switch closed $t > 90$ sec.," "Timer contacts closed $t > 90$ sec.," and "Power supply functions $t > 90$ sec." Figure 29.17 shows these tree levels.

Analyzing "Timer contacts closed $t > 90$ sec.," the logic indicates a component-related event. Thus an OR gate with primary, secondary, and command events are required as causes. The primary internal failure is "Timer contacts fail closed." The command event must be further analyzed. It is seen that the timer relay commands the contacts closed. This could be caused by failure of power to the relay since the

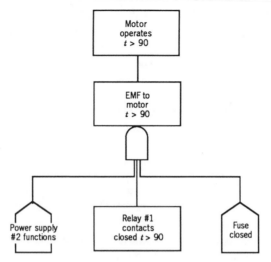

Figure 29.16   System fault tree, top levels.

relay opens the contacts when under power, the spring loaded switch being open or the timer function failing. Figure 29.18 shows these final tree levels.

The analyst must note at this point, that the End events of Figure 29.18 that will cause power loss, will result in Relay 1 opening and thus prevent the Top event. Therefore the tree of Figure 29.18 should be correctly synthesized as having only the single End causing "Timer contacts command closed". This event should be "Timer function in relay fails."

29.3   CUT SET DEVELOPMENT

In previous sections, we have explored the logical process of fault tree synthesis. We shoud recall that the fault tree represents an accident system or sets of events in Boolean logic form that lead to an accident. Having constructed this model of an accident process, we must exercise or analyze it for useful information as to the state of safety of the system. Although it is true that an examination of the tree structure may provide some information about the state of the system, the principal reason for synthesizing the tree is to find the minimal cut sets. And from these cut sets quantitative or qualitative analysis may be performed on the system.

We have previously developed the algebra of dichotomous events called Boolean algebra. We have also examined the nature of cut sets as applied to reliability systems. We should recall that a minimal cut set is a nonredundant set of events that will cut a reliability system. Approached from the safety viewpoint, we may consider a minimal cut set as a nonredundant set of events that *cause* an undesired accident event. In the reliability case, cutting the system with a cut set caused a failure of the reliability system. In the case of a fault tree, which is an accident system, a minimal cut set will be a set of events that causes the Top event of the tree. In this sense, it is logically a path, although we will retain the name *cut set* for a set of events that cause an undesired state in the system.

A cut set is made up of the End events of the fault tree. It is created by the logical relationships between these events as formed by the Boolean algebra logic gates through which the End events must act to reach the Top event. We have previously examined symbols for gates. In this development we will deal only with simple AND and OR gates, showing how they correspond to Boolean operators of *intersection and union*.

29.3.1   OR Gate

The fault tree symbol OR represents the Boolean *union* of events entering the gate to pass the fault through the gate. Thus, the OR gate logic is that denoted by the Boolean symbol ∪. To write this Boolean logic in the determination of cut sets

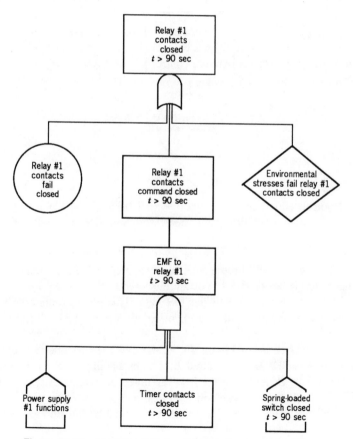

Figure 29.17   Analysis of "Relay #1 contacts closed $t > 90$ sec."

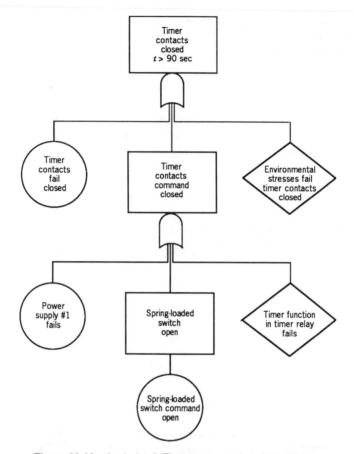

Figure 29.18   Analysis of "Timer contacts closed $t > 90$ sec."

through a Boolean equation, the symbol + will be used. The plus sign is used because the arithmetic operation performed through an OR gate, when the events of the gate are associated with probabilities, is addition. From our previous section on Boolean algebra, we know that simple addition will not always provide a precise answer, but in most cases will provide an acceptable conservative bound to the true value if that value is the probability of an undesired event.

Figure 29.19 illustrates the fault tree OR gate. This Figure represents an event in which either contact 1 or contact 2 open would interrupt EMF to the main DC bus.

Examining the example of Figure 29.19, we can see that one or more of the events A, B, or C will cause the output E. The probability of event E, given the probability spaces A, B, and C, is the Boolean union of these events. Figure 29.20 illustrates the union of these events in the most general case.

The probability space of event E can be represented by the Boolean equation:

$$E = A \cup B \cup C \qquad (29.1)$$

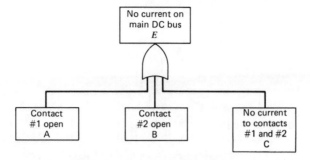

Figure 29.19 Fault tree OR gate.

If we substitute a + sign for the Boolean union symbol, the equation may be rewritten.

$$E = A + B + C \qquad (29.2)$$

Equation (29.1) represents the precise probability or the precise area of the union as shown in Figure 29.20. This equation implies that the intersections of the Boolean algebra will be properly treated so that the true value emerges. If we perform the simple addition as shown in equation (29.2), we will get too large a value for the probability if there are intersections. We should recall that the true value of the event E, if Figure 29.20 accurately depicts the conjunction of the three events, is the following.

$$E = A + B + C - AB - AC - BC + ABC \qquad (29.3)$$

where AB, AC, BC, and ABC represent intersections of the events A, B, and C.

Thus, we can see that, in this case, equation (29.2) represents an upper bound on the true value. If the events are mutually exclusive, the equation will be precise. If the events are of reasonably small probabilities, as we would hope in fault events, equation (29.2) will provide a close upper bound to the the true value.

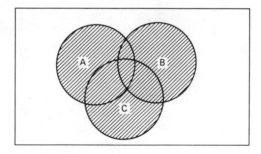

Figure 29.20 Venn diagram of "No EMF to main DC bus."

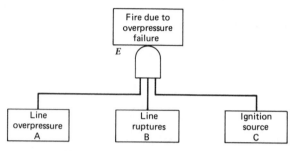

Figure 29.21   Fault tree AND gate.

29.3.2   AND Gate

The fault tree symbol AND represents the Boolean *intersection* of events passing through the gate. Thus, the AND gate logic is denoted by the Boolean symbol ∩. To write this logic in the determination of cut sets from a Boolean equation, algebraic multiplication symbols will be used. Conveniently, this is not a symbol but is merely showing the event symbols adjacent to each other. The multiplication symbol is used because the arithmetic operation used to approximate an intersection of events is multiplication when the events are associated with probabilities. We also know from our brief study of Boolean algebra, that such a multiplication process is only precise if the events in question are *independent*. If they are not independent, the multiplication process cannot be relied upon to provide a conservation bound as in the OR gate case. However, if the probabilities are small, as they should be if they are fault events in accident systems, multiplication will provide quite a good approximation of the true value. Figure 29.21 illustrates an AND gate.

The Venn diagram of Figure 29.21 is shown in Figure 29.22.

Examining Figure 29.21, we can see that the output E will occur only if all of the fault events A, B, and C occur. Figure 29.22 is a Venn diagram of the intersection of all three events. It represents the true intersection. The intersection shown in Figure 29.22 can be represented by a Boolean equation.

$$E = A \cap B \cap C \tag{29.4}$$

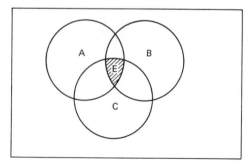

Figure 29.22   Venn diagram of fire due to overpressure failure.

Equation (29.4) represents the true value of the intersection. For computation purposes, we will replace the intersection symbols with multiplication.

$$E = (A)(B)(C) \quad \text{or} \quad E = ABC \qquad (29.5)$$

Equation (29.5) will, of course, be an approximation of the true value if the events A, B, and C are not independent.

29.3.3 Boolean Equation

Having examined the logical elements of the fault tree, we are prepared to write the Boolean equation of a tree. We will use the top-down approach to write the equation for the synthesized tree of Figure 29.23.

Starting with the Top event T, we can come down through the top OR gate and begin to write the equation.

$$T = E1 + E2 + E3$$

Proceeding to the next level we can write:

$$T = A + B + C + (E21)(E22) + (E31)(DF)$$

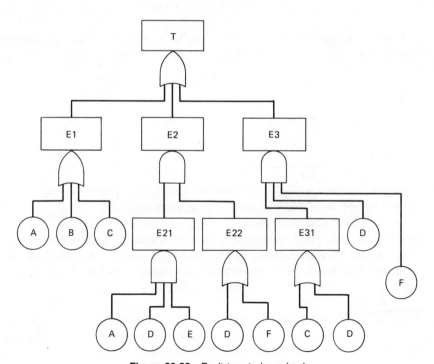

Figure 29.23 Fault tree to be solved.

Proceeding down to the final End event level we can write:

$$T = A + B + C + (ADE)(D + F) + (C + D)DF$$

Applying the laws of Boolean algebra to the foregoing equation we can begin to simplify.,

$$T = A + B + C + ADE + ADEF + CDF + DF$$

Further simplifying:

$$T = A + B + C + ADE(1 + F) + DF(C + 1)$$

Noting that:

$$1 = \text{universe} \quad \text{and} \quad Z \cap 1 = Z$$

$$T = A + B + C + ADE + DF$$

Continuing to further simplify:

$$T = A + B + C + DF$$

This final expression is the Boolean equivalent expression for the original tree of Figure 29.23. The minimal cut sets of this tree are then:

$$C_1 = A \qquad C_3 = C$$

$$C_2 = B \qquad C_4 = DF$$

We can then observe that if event A occurs, the Top event occurs. Likewise, we can observe the same for events B, C, or D and F. This tree or this system has three order 1 cut sets and one order 2 cut set. This system is quite hazardous and clearly not acceptable if the Top event is of substantial severity.

If we are considering safety countermeasures for this system but know nothing of the relative likelihoods of events A, B, C, D, or F, we could observe the following. Events A, B, and C are of equal Importance in causing T and are the most Important. If we know that event T has occurred, it is equal-likely that A, B, or C were involved. Events D and F are of lesser Importance, given that they are equal-likely with the other three events.

There are other approaches to manually solving for minimal cut sets. There is a bottom-up approach which proceeds in a similar manner to the top-down approach above. There is also an algorithm called MOCUS. The MOCUS approach is useful for computer programs in that it is a more systematic method and can easily be programmed. The MOCUS algorithm can be found in any of a number of good reference works.

29.3.4   Boolean Equivalent Tree

From the minimal cut sets of a tree, a new tree may be synthesized. This tree will be logically equivalent to the original tree. We may think of this new tree as the *pruned* version of the original tree. It is properly called the *Boolean equivalent tree*, shown in Figure 29.24.

If management or the analyst wishes to examine a tree for the state of safety of a system, they need not examine the original tree. The Boolean equivalent tree will be quite satisfactory. If a tree has thousands of cut sets the Boolean equivalent itself may be quite complex and provide little insight to the state of safety.

29.3.5   Path Sets

A path set of a fault tree can be defined as the nonredundant set of events that will prevent the occurrence of the Top event. Considered logically, this means that *each* path set must cut all of the cut sets, since each path set must prevent all cut sets from faulting. Thus, in the example above, we can quite simply find the path sets as:

$$P_1 = \overline{A}, \overline{B}, \overline{C}, \overline{D}$$

$$P_2 = \overline{A}, \overline{B}, \overline{C}, \overline{F}$$

Note that these events are the complement of the original events. Thus, they are the nonoccurrence of the original End events of the tree.

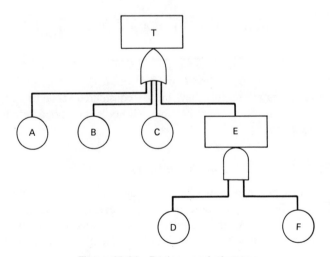

Figure 29.24   Boolean equivalent tree.

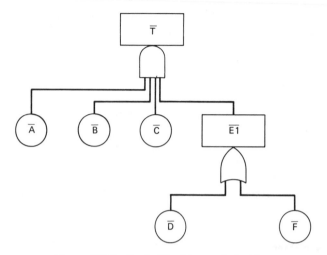

Figure 29.25 Dual of Boolean equivalent tree.

Of course, if there are thousands of cut sets a path set solution by inspection would not be possible. A structured solution for path sets can be developed from the original tree. A dual of the synthesized tree is created, and the cut sets of the dual will represent the path sets of the original tree. A tree dual is created by changing OR gates to AND and AND gates to OR and making each event in the tree its own complement. A simpler approach is to find the dual of the Boolean equivalent tree. The cut sets of this dual will also be the path sets of the original tree. In this example we can synthesize a dual of the Boolean equivalent tree of Figure 29.24.

Writing the Boolean expression for Figure 29.25:

$$T = (\overline{A}\ \overline{B}\ \overline{C})(\overline{D} + \overline{F}) = \overline{A}\ \overline{B}\ \overline{C}\ \overline{D} + \overline{A}\ \overline{B}\ \overline{C}\ \overline{F}$$

If we examine the original tree we can see by inspection that the Top event will be prevented if events A, B, C, and D or F do not occur.

Although path sets are useful, they will be considered secondary in the safety analysis of the system. Their quantification will not be described herein, although it is possible to do so.

29.3.6 Fault Tree and Network Methods

A fault tree is an accident system. Thus, the logic of the tree transformed directly into a series–parallel system would be a set of events leading to an undesired outcome, or an accident. As an example, we can examine Figure 29.26.

If we transform this tree to a series–parallel system it would appear as Figure 29.27.

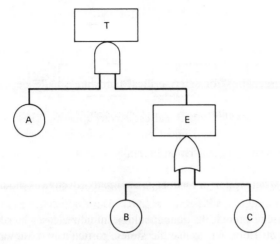

Figure 29.26 Basic fault tree.

If we find the cut sets of Figure 29.26 they will be:

$$C_1 = A, B$$

$$C_2 = A, C$$

The path sets of Figure 29.26 will be:

$$P_1 = \overline{A}$$

$$P_2 = \overline{B}\,\overline{C}$$

The path sets of Figure 29.27 will be the same as the cut sets of the fault tree since the cuts of tree assure the output event occurs.

If we create a dual of an accident system, we will have a reliability system. Thus, the dual of Figure 29.27 will appear as Figure 29.28.

The cut sets of Figure 29.28 will be the cuts of the original tree since to cut a reliability system will give an undesired event. The path sets of Figure 29.28 also are the paths of the original tree. It can be seen that this is the case.

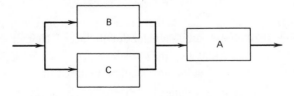

Figure 29.27 Series–parallel system of fault tree.

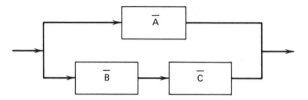

Figure 29.28   Reliability system of fault tree.

29.4   FAULT TREE QUANTIFICATION

A final and objective method of evaluating the acceptability of the state of safety of a system is through quantifiable measures. Development and analysis of the cut sets is a necessary prelude to quantification. In some cases, qualitative analysis may be sufficient. However, in many systems, particularly those for which the Top events are catastrophic and involve very large losses, a quantified measure of the Top event is necessary to make a proper judgment as to acceptability.

In this section, we examine a number of quantifiable parameters in a fault tree. Some of these parameters are input values used to compute the final measures of safety performance.

The principal measures of the Top event or of any major subsystem node within the tree are the following:

Unreliability
Unavailability
Expected number

Measures that are principally used as input values, but may also be used to measure Top events or nodes are the following:

Fault rate
Rate of fault
Mean down time

A measure used to judge which End events are most involved in the causation of the Top event or node is:

Importance

A measure that is useful to judge the acceptability of a standby system is:

Undependability

It is apparent that each of these parameters is a negative measure of performance. As the values of these parameters increase, the situation becomes less desirable.

Two of the principal measures of the Top event may also be expressed in complementary values. These would be reliability and availability. However, our position in safety will be to continue to describe undesirable factor levels.

In the descriptions of the measures of safety performance that follow, the author is indebted to Dr. Jerry Fussell and Dr. Howard Lambert, who have done much theoretical work in this field, completing the vital link between theory and application.

29.4.1 Fault Rate

A fault rate, λ_i, may be defined such that it is the instantaneous rate of fault of an event or failure of a component. This implies that λ is the probability of the event fault or failure in the interval $t + \Delta t$, given that it has not occurred in the interval 0 to t. In general, λ is a function of t. Therefore, the fault rate should be properly related with the (t) argument as $\lambda_i(t)$, where i implies a rate for a particular event. Although the fault rate does vary with t, and thus all of the quantitative descriptors of a fault tree also vary with t, for our manual computations we will assume λ invariant with t and drop the t argument for this parameter and all others on which it depends.

The notations for fault rate for the three levels of events within a tree are the following:

Fault rate: λ_i (End event), Λ_k (cut set), Λ_T (Top event)

The fault rate of an event or component must be an input to the quantitative analysis of a fault tree. Fault rates may be derived from rigorous testing of a component or human function or be estimated from surrogate data. The reliability group is the most convenient source of fault rate data in an industrial organization. There are national data banks that contain fault or failure rates of many classes of components.

If one desires to determine the fault rate of cut sets or End events, they may be hand calculated as shown in equations (29.6) and (29.7).

$$\Lambda_k \cong Rf_k/A_k \qquad (29.6)$$

$$\Lambda_T \leqslant \Sigma\Lambda_k \qquad (29.7)$$

If fault events can be repaired after going to the fault or failed state, we must consider a measure of the likelihood of this repair as it will have significant influence on the likelihood of faulting of a cut set and therefore the Top event.

29.4.2 Mean Down Time

We can define a repair rate, $\mu_i(t)$, such that it is the instantaneous rate of repair at a given value of t. We will assume that μ_i is constant with t. The reciprocal of μ, is the mean time between repairs, τ.

$$\tau = 1/\mu_i \qquad (29.8)$$

This measure is sometimes called the *mean down time*. This value is the mean time a component or other fault End event remains in the fault or failed state before it is brought to a nonfault state by repair action. Here, we should note that if a component is failed by secondary cause, it will not be repaired by repair of the component alone. An electrical component that fails due to an overvoltage condition is not repaired by replacement of the component if the overvoltage condition remains. We can call this type of system statistically noncoherent. Trees that are statistically noncoherent can be transformed to statistically coherent systems.

Before proceeding to develop the equations for the quantifiable elements of a fault tree, we should define the notation that will be used for these events. In this notation we will assume λ and τ are constant with t and drop the t argument.

Unavailability: $\bar{a}_i$ (End event), $\bar{A}_k$ (cut set), $\bar{A}_T$ (Top event)
Unreliability: $\bar{r}_i$ (End event), $\bar{R}_k$ (cut set), $\bar{R}_T$ (Top event)
Rate of fault: rf_i (End event), Rf_k (cut set), Rf_T (Top event)
Expected number of faults: Ef_i (End event), Ef_k (cut set) Ef_T (Top event)
Importance: I_e (End event), I_k (cut set)
Undependability: d_i (component), D_i (system or subsystem)

29.4.3   Unreliability

Unreliability of an End event, $\bar{r}_i$, may be defined as the probability that at least one fault occurred in the interval 0 to t given that it was not faulted or failed at $t = 0$.

The unreliability of a cut set, $\bar{R}_k$, is the probability that all events in the cut set were in the fault state for at least one instant, given that no events had faulted prior to $t = 0$. Unreliability of an event does not consider repairs as possible. The unreliable state occurs when the event first faults. In the case of unreliability of a cut set, repairs of the events will alter the unreliability. Consider Figure 29.29.

This figure examines the transition of a two-event cut set to an unreliable state. Neither event has faulted at state 0. In state 2 the second event has faulted. In state 3, the second event has been repaired and the first event has faulted. We assume that the second event was repaired before the first event faulted. The cut set remains in a reliable state at state 3 due to repair. Without repair the cut set would be in

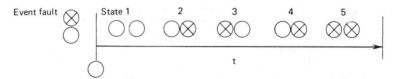

Figure 29.29   Two-event cut set transition to unreliability with repair.

the fault state at state 3. In state 4, the first event is repaired and the second event has again faulted. Finally, in state 5, the first event has faulted prior to repair of the second event. This causes an unreliable state. The difference in the likelihood of the cut set faulting with and without repair is quite clear.

The unreliability of the Top event, $\overline{R}_T$, is the probability that the Top event occurs at least once during the interval 0 to t, given that it has not occurred at $t = 0$.

Assuming λ invariant with t, the unreliability of an event is the following:

$$\overline{r}_i = 1 - e^{-\lambda_i t} \tag{29.9}$$

As may be recalled from the Poisson distribution, there is quite a good approximation of $\overline{r}_i$ if $\lambda_i t$ is small. We will adopt the standard of .01. Thus, the following equation obtains:

$$\overline{r}_i \leq \lambda_i t \quad \text{if} \quad \lambda_i t \leq .01 \tag{29.10}$$

Recall from the earlier section on cut sets that the probability of a cut set is the product of the probabilities of the events of the cut sets. Therefore, assuming no repair, the unreliability of a cut set is the following:

$$\overline{R}_k = \prod^n r_i \tag{29.11}$$

where n = number of events in the cut set.

In general, the unreliability of a cut set is equal to the probability of at least one cut set fault in the interval 0 to t.

$$\overline{R}_k = 1 - e^{-\Lambda_k t} \tag{29.12}$$

Assuming Λ is invariant with t and with repair, if repair is possible.

The calculation of Λ_k with repair is cumbersome as a direct approach to cut set unreliability so we will examine the more indirect approach. Equation (29.29) is the approach to $\overline{R}_k$.

$$\overline{R}_k \leq Ef_k \tag{29.13}$$

The expected number of faults of a cut set with repair is:

$$Ef_k = \int Rf_k \tag{29.14}$$

The integral of the rate of fault of a cut set, given that it is constant with t, is given by a set of equations in Section 29.4.5.

The unreliability of the Top event will be the union of the cut sets. If there are no intersections, the unreliability of the Top will be the sum of the unreliabilities of the cut sets. If there are intersections between cut sets the sum will provide a conservative upper bound on the unreliability of the Top event.

$$\overline{R}_T \leq \sum^{N} \overline{R}_k \tag{29.15}$$

where N = number of cut sets.

In most cases with reasonably small probabilities of events, this conservative upper bound will be quite close and most satisfactory. However, if the unreliabilities of the cut sets are large values and thus add to an unacceptably large number ($>.4$), the intersections are large and must be accounted for. In that case, equation (29.16) is suggested as an alternate calculation.

$$\overline{R}_T = 1 - \prod^{N} R_k \tag{29.16}$$

Equations (29.15) and (29.16) are valid with or without repair.

29.4.4   Unavailability

The *unavailability* of an End event, $\overline{a}_i$, may be defined as the probability an event is in the fault state *at t* given faults and repairs that have occurred prior to t. There is a further assumption that the event was in the nonfault state at $t = 0$. If the events are unrepairable, the unavailability of an event is equal to the unreliability.

The unavailability of a cut set, $\overline{A}_k$, is defined as the probability that the cut set is in the fault state at t considering faults and repairs made to events of the cut set prior to t and further assuming that all events in the cut set were in the nonfault state at $t = 0$. The unavailability of the Top event, $\overline{A}_T$, is the probability that the Top event is in occurrence at t considering event faults and repairs within the cut sets prior to t. This value also assumes that all events in all cut sets were in the nonfault state at $t = 0$. If all events are unrepairable, unavailability of the cut sets and the Top event are equal to their corresponding unreliability.

The unavailability of an event is dependent upon both the fault rate and mean down time and takes the form shown in Figure 29.30.

The unavailability of an event is composed of the two parts, the asymptotic value and the variable value. From Markov analysis of the transition of the event from

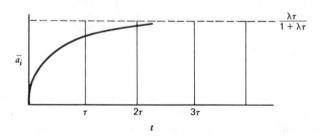

Figure 29.30   Event unavailability.

the normal to fault state, the probability that an event will be in a fault state at t, given λ and τ, is equation (29.17).

$$\bar{a}_i = [\lambda_i\tau_i/(1 + \lambda_i\tau_i)][1 - e^{-(\lambda_i + 1/\tau_i)t}] \qquad (29.17)$$

The asymptotic part of equation (29.17) is $\lambda\tau/(1 + \lambda\tau)$.

As the value of the variable part of unavailability approaches 1.0, unavailability approaches its asymptotic value. For manual calculations, we may assume the variable part of the equation need be $>.9$ for this to take place.

If $\lambda\tau$ is small, $\leq .1$, the asymptotic portion of unavailability can be approximated by $\lambda\tau$ itself.

$$\bar{a}_i = \lambda_i\tau_i \qquad \text{if } \lambda_i\tau_i < .1 \qquad (29.18)$$

The unavailability of a cut set is quite simply the product of the unavailabilities of the events of the cut set. Recall again that if there is no repair, cut set unavailability is equal to unreliability.

$$\bar{A}_k = \prod^{n}\bar{a}_i \qquad (29.19)$$

The unavailability of the Top event is equal to the union of the unavailabilities of the cut sets. This can be conservatively bounded by the sum of the cut set unavailabilities.

$$\bar{A}_T \leq \sum^{N}\bar{A}_k \qquad (29.20)$$

29.4.5 Rate of Fault

The *rate of fault* is defined as the expected number of faults of an event, cut set, or Top event, per unit time at time t. Although it sounds similar to fault rate, recall that fault rate is an instantaneous probability of faulting given that it had not previously faulted. The rate of fault of an event is normally less than the fault rate because an event cannot fault if it faulted previously and not repaired.

The rate of fault of a nonrepairable event is:

$$rf_i = r_i\lambda_i \qquad (29.21)$$

It is clear that in the nonrepairable case the event rate of fault must be the product of event reliability and the instantaneous fault rate.

In a like manner, if the event is repairable:

$$rf_i = a_i\lambda_i = (1 - \bar{a}_i)\lambda_i \qquad (29.22)$$

Recall again that these equations consider λ constant with time.

We can see that for events, the rate of fault is less than the fault rate by the factor of unreliability for unrepairable events, and unavailability for repairable events.

We will take greater care in developing the expression for rate of fault for cut sets. This expression will form the basis for calculating unreliability of a cut set. Also, the expression in its final form is not obvious.

We will develop the expression for a two-event cut set going through a transition from two events in a nonfault state to both events in a fault state and thus the cut set in the fault state. Figure 29.31 illustrates the transition states for this case.

The two-event minimal cut set may enter the fault state through one of three possible transitions. The cut set may go directly from I to IV by both events entering the fault state. If the events are repairable, equation (29.21) will describe that transition.

$$a_1\lambda_1 a_2\lambda_2 \tag{29.23}$$

The transition from state II to IV is the following rf_i.

$$\bar{a}_1 a_2\lambda_2 \tag{29.24}$$

The transition from state III to IV is the following rf_i.

$$a_1\bar{a}_2\lambda_1 \tag{29.25}$$

The total expression for the rate of fault of a two-event cut set considering λ as a function of t is the following:

$$Rf_k = a_1 a_2\lambda_1(t)\lambda_2(t)dt^2 + \bar{a}_1 a_2\lambda_2(t)dt + \bar{a}_2 a_1\lambda_1(t)dt \tag{29.26}$$

Noting that dt^2 is approximately zero, we have:

$$Rf_k = \bar{a}_1 a_2\lambda_2(t)dt + \bar{a}_2 a_1\lambda_1(t)dt \tag{29.27}$$

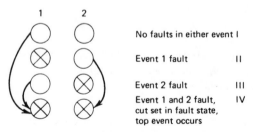

No faults in either event	I
Event 1 fault	II
Event 2 fault	III
Event 1 and 2 fault, cut set in fault state, top event occurs	IV

Figure 29.31   Transition states for two-event cut set.

In performing hand calculations, we may further simplify.

$$a_1, a_2 \cong 1.00 \tag{29.28}$$

Assuming that λs are constant with t, the following expression will provide a conservative bound on the rate of fault for a two-event cut set, assuming repair.

$$Rf_k \leq \bar{a}_1\lambda_2 + \bar{a}_2\lambda_1 \tag{29.29}$$

For cut sets of higher order, the rate of fault will be every combination of unavailability and fault rate.

In an analogous fashion the rate of fault for a two-event cut set without repair is:

$$Rf_k = \bar{r}_1\lambda_2 + \bar{r}_2\lambda_1 \tag{29.30}$$

The rate of fault, Rf_T is conservatively bounded by the sum of the rate of faults of the cut sets.

$$Rf_T \leq \Sigma Rf_k \tag{29.31}$$

29.4.6 Expected Number of Faults

The *expected number of faults* of an End event, Ef_i, may be defined as the mean number of fault events that will occur in the interval 0 to t, assuming all events were in the nonfault state at $t = 0$.

The expected number of faults of a cut set, Ef_k, is the mean number of times all events of the cut set will be in a fault state at the same instant during the interval 0 to t, assuming all events of the cut set were in the nonfault state at $t = 0$. The expected number of faults of a Top event, Ef_T, is the mean number of times the Top event will occur in the interval 0 to t assuming all events are in the nonfault state at $t = 0$.

As the rate of fault of an event is the instantaneous expected number of faults of an event, the expected number of faults over an interval t, is simply the integral of the event rate of fault over the interval.

$$Ef_i = \int_0^t rf_i(t')dt' \tag{29.32}$$

If λ is invariant with t, the rate of fault is also constant. The rate of fault of an event is then simply:

$$Ef_i = rf_i(t) \tag{29.33}$$

The expected number of faults of a cut set is equal to the integral of the rate of fault of the cut set.

$$Ef_k = \int Rf_k \qquad (29.34)$$

If the rate of fault of the cut set is invariant with t because Λ_k is invariant with t, then the expected number of faults of a cut set is:

$$Ef_k = Rf_k(t) \qquad (29.35)$$

The expected number of faults of a Top event is the Boolean union of the expected number of faults of the cut sets. This can be closely and conservatively bounded, if the expected number is low, by the sum of the expected number of faults of the cut sets.

$$Ef_T = \leq \sum_{}^{N} Ef_k$$

29.4.7 Cut Set Importance

Importance is the measure of the sensitivity of the next element above in the hierarchy of the tree to the faulting of that element. Minimal *cut set Importance*, I_k, can be defined as the probability that the cut set is in a fault state, given that the Top event has occurred. Importance can be considered a weighting factor by which to judge the most effective point of attack to reduce the likelihood of the Top event.

If we assume that F_k is a minimal cut set in a fault state and that F_T is an occurrence of the Top event, then:

$$I_k = P[F_k|F_T] \qquad \text{(assuming } \lambda \text{ invariant with } t) \qquad (29.36)$$

where $P[F_k|F_T]$ is conditional probability, $P[F_k]$ given $P[F_T]$.

$$P[F_k|F_T] = \frac{P[F_k \cap F_T]}{P[F_T]} = \frac{P[F_k]P[F_T|F_k]}{P[F_T]} \qquad (29.37)$$

However:

$$P[F_T|F_k] = 1.00$$

Therefore:

$$I_k = \frac{P[F_k]}{P[F_T]} \qquad (29.38)$$

Equation (29.38) says simply that the Importance of a cut set is a ratio of the probability of the cut set fault occurrence to the probability of the Top event occurrence. Note that since the sum of the cut sets probabilities are approximately equal to the probability of the Top event, the sum of the cut set Importances must equal 1.00.

Importances may be calculated using either unreliability or unavailability. Unavailability may be used to calculate Importance if the events are repairable and the Top event is noncatastrophic or a single occurrence of the Top event does not preclude repair and recovery of system function. In addition, unavailability should be used as the basis for Importance, when there is concern for the nonoccurrence status of the Top event at a point in time. Such would be the case for a long-range radar that need only be up when there is a target. Conversely, unreliability should be used to calculate Importance if a single occurrence of the Top event damages the system beyond repair or moves the system to a nonrecoverable state. This would be the case of an inadvertent launch of a missile.

29.4.8   Event Importance

Event Importance may be defined as the probability that a specified event is contributing to the Top event occurrence, given that the Top event has occurred. This means that the event in question has faulted and is in a cut set that has faulted. This form of Importance is generally attributed to Fussell-Vesely.

If we assume that F_K is at least one cut set containing the event e in a fault state and F_e is event e in a fault state, then:

$$I_e = P[(F_K \cap F_e)|F_T] = \frac{P[(F_K \cap F_e) \cap F_T]}{P[F_T]} \qquad (29.39)$$

$$= \frac{P[F_K \cap F_e][P(F_T|F_K \cap F_e)]}{P[F_T]} \qquad (29.40)$$

However:

$$P[F_T|(F_K \cap F_e)] = 1.00 \qquad (29.41)$$

Therefore:

$$I_e = \frac{P(F_K \cap F_e)}{P[F_T]} = \frac{P[F_K][P(F_e|F_K)]}{P[F_T]} \qquad (29.42)$$

However:

$$P(F_e|F_K) = 1.00 \qquad (29.43)$$

Therefore:

$$I_e = \frac{P[F_K]}{P[F_T]} \tag{29.44}$$

The $P[F_K]$ contains the probability space of all of the minimal cut sets containing e being in a fault state. Therefore:

$$I_e = \frac{P(f_1 \cap f_2 \cap \ldots f_n)}{P[F_T]} \tag{29.45}$$

where f_i is the event that the ith cut containing e is in a fault state and n is the number of cut sets containing e.

Equation (29.45) cannot be evaluated exactly but can be closely approximated by equation (29.46).

$$I_e \cong \sum^{N_e} I_{ke} \tag{29.46}$$

where

$N_e =$ the number of cut sets containing e
$I_{ke} =$ the importance of the cut sets containing e

Equation (29.46) states that a basic event Importance is approximately equal to the sum of the Importances of the cut sets containing that event.

29.4.9   Order Importance

A frequently voiced objection to fault tree analysis is that without input data the results of the analysis are not useful. It is true that, on most complex systems, event rates and repair measures may not be available for all End events. If only a few End-event data are missing, conservative estimates may be made of the missing values and Importance calculations combined with sensitivity analysis will quickly show if the missing values can be estimated without losing analysis validity. However, if the majority of the End-event data are unknown, another approach can be most useful in determining which End events should be targeted for countermeasures.

Importance of End events may be calculated with arbitrarily assigned End-event probabilities of equal values. If, in a large tree, one ranks the top 20 End events by Importance calculated by assigned End-event probabilities and also by rigorously determined rates and repair values, a remarkable correlation will be found between the two rankings. This would suggest that if the purpose of fault tree analysis is to examine the causation sets of the Top event (cut sets) and determine which End events would be the most vulnerable to improvement with countermeasures, accurate End event fault rates and mean down times are not necessary. This is true because the primary determinates of Importance are the order of the cut sets in which an event appears as well as the number of cut sets in which an event appears.

Importance determined with assigned End event values is called *order Importance*.

There remains the question of what values to assign. There is some warping of Importance as the assigned values move from likely to less likely, although not of significant impact. It is suggested that an intermediate level of rate ($\sim E-3$), with no repair would provide satisfactory indications of the relative Importances of critical End events.

If the fault tree contains some End events that are obviously of greater likelihood than others, it is suggested these events be separated by one to two orders of magnitude. Thus, if the fault tree contains very reliable component End events and quite unreliable human End events, the components might be assigned rates of $\sim E-4$ while the human functional failures are assigned rates of $\sim E-2$.

This form of Importance is called *relative order Importance*.

29.4.10   Undependability

Undependability is a term that may be applied to a system, subsystem, or component that is in a standby mode during normal system function. *Undependability* is defined as the probability that the standby unit will come on line when needed and function for the required period of demand. Undependability is the complement of dependability.

We can understand the concept if we consider an emergency power supply that is to serve when the primary source of power fails. The undependability of this power supply is the probability that the power supply will either *not* come on line when needed or will *not* run for the required period of demand.

The undependability of a system element is thus the Boolean union of two general sets of cut sets. One set is the set that encompasses the cut sets that will prevent the standby system coming on line on demand. The other encompasses the cut sets that will prevent the standby system functioning for required period t.

Implicit in undependability is the notion that a component may fail while in standby. If a system may fail while in standby or storage, it may also be inspected and repaired while in standby. The failure to come on line on demand may be described by unavailability, of if not repairable or not inspected while in standby, unreliability. Failure of the system to function for the period of demand will be simply unreliability. These two measures will contain different cut sets.

Undependability may be described by equation (29.47).

$$\overline{D}_i = 1 - (a_i(s))(r_i(t)) \tag{29.47}$$

where

$$a_i = \text{the availability for period of storage}$$

$$r_i = \text{the reliability for the period of demand}$$

29.4.11 Common Cause Analysis

A most important task in the analysis of fault tree cut sets is that of determining possible environmental causes of fault or failure events within the system. Such environmental causes of faults have the disturbing possibility of turning faults that were treated as independent in the determination of cut sets and subsequent quantification, into dependent events that occur when an environmental effect occurs. Thus, such causes of fault, called *common cause*, have the effect of modifying minimal cut sets and the quantification of the tree. For instance we may have designed into a system reduntant components to assure needed reliability or safety only to find that the likelihood of failure of the two events is only slightly greater than that of a single event. Even worse, if the common cause is itself likely, the redundant system may be more likely to fail than a single event without the environmental or common cause of failure. A common cause is sometimes called a *common killer*.

Consider a system and its minimal cut sets.

$$C_1 = A, B, C$$

$$C_2 = A, B, D, E$$

$$C_3 = A, B, E, F, G$$

Assume that event B is a high probability event, perhaps a human failure. It is common to all cut sets but is protected by other events of low probability. Assume further that there is a common cause that affects events A, C, and E such that the occurrence of the common cause causes events A, C, and E to occur.

$$CC_1 \rightarrow A, C, E$$

Events A, C, and E are called *common-mode* events for the common cause CC_1. It need not be a certainty that the occurrence of the common cause will cause all of the common-mode events. If the probability is relatively high it must be considered as if a certainty. If we apply the common cause to the three cut sets, we have a new set of cut sets.

$$C_1 = CC_1, B$$

$$C_2 = CC_1, B, D$$

$$C_3 = CC_1, B, F, G$$

We can readily see that cut sets C_2, and C_3 are nonminimal. The new minimal cut set, considering the common cause is:

$$C_1 = CC_1, B$$

Thus, what was initially a system with three cut sets of minimum order three becomes a system of one cut set consisting of one event that may be of high probability. If this is so, we have essentially a single-point fault system.

A short list of common causes would be the following:

Vibration

Temperature

Ionizing radiation

Pressure

Corrosion

Metal fatigue

Common test procedures or equipment

Manufacturing defect (common vendor)

If one or more of these causes is normal rather than abnormal, the system must be carefully examined for acceptability.

29.4.12   Fault Tree Computer-aided Analysis

As can readily be seen, fault tree quantitative calculations are complex and time consuming. Because of their complexity, the analyst is never quite sure if the values derived are correct. This time-consuming complexity and uncertainty mitigate against fault tree quantification and thus against the entire fault tree process. Furthermore, if the tree has more than 50 or 60 cut sets, manual calculations become impractical. Computer programs are the obvious solution. There are several available.

Computer calculation of the fault tree values facilitates a variety and depth of analysis that would be out of the question for manual calculations. For instance, the analyst can make calculations for a number of exposure values t. These can be done in equal intervals or at selected intervals, examining the effect on Top event measures such as unavailability or expected number. If the analyst is uncertain of the validity of event rates and mean down times, he or she can vary these in sensitivity studies of Top event measures, subsystem measures, or even event measures such as Importance.

Computer analysis of fault trees has become an essential part of the process. An analyst who contemplates fault tree analysis should, as a matter of course, acquire a computer program adequate for his or her needs and become familiar with its use. To date, the majority of fault tree analytical programs have been written for the mainframe computer. These large machines have the advantage of rapidity of computation coupled with virtually unlimited memory size to handle the large amounts of data that major systems engender. However, their use has several major deficiencies.

Lack of access is a major detriment to the use of mainframe computers. Such computers are normally controlled by a special division within an organization. Use of these valuable and costly assets is limited by account numbers and user passwords. This provides a minor barrier to occasional use. Furthermore, the computer and its access terminals may be remote from the work area of the analyst. The analyst must move to a new work site when using the mainframe or transmit the problem to a computer specialist who types in the data, operates the machine, and returns with the results. The deficiencies of such a system are obvious.

If the analyst is involved in a major fault tree project that will take some months, such as analysis of a nuclear power facility, he or she can move to the new work site on a semipermanent basis and devote himself or herself to the computer task. However, most fault tree analysis is done by the system safety analyst on an intermittent basis. The analyst has other tasks beyond fault trees.

The use of the mainframe computer programs that the author has seen, require a period of training. Data input requires precise coding for the many options available. For the normal analyst who needs such computational power only occasionally, some retraining will be required on each use. Even in large organizations that maintain computer service personnel to assist the analyst, these people will also need retraining to reacquaint themselves with the intricacies of fault tree data input before they can provide service to the analyst.

Such difficulties in the use of mainframe programs discourage their use except for the most complex and important problems where the investment of the time and effort are warranted.

These problems may be summarized with a single word: *control*. For the analyst to perform quantitative fault tree analysis whenever needed, that person must have easy, direct control of the computer and the program. This control is essential if this flexible and powerful analytical tool is to be fully utilized. Thus, the fault tree analyst should have convenient access in the form of a microcomputer or a terminal connected to mainframe or minicomputer at his or her desk or nearby. The analytical software should be menu driven to minimize training.

Menu-driven software implies that the user need only answer questions to input data and activate the programs' computing features. The menu should be easily understood with a minimum of training. Suitable software should have a most legible output with values organized and clearly labeled in English as to the variables being displayed.

It should not be necessary to enter the tree logic more than once. This implies that once the tree logic is entered, the program should be capable of finding minimal cut sets, calculating all quantitative parameters with addition of event rates and mean down times, and plotting the tree. It would be helpful if the program could be edited following initial data input.

Analysts should endeavor to procure programs with these characteristics. One such program, called TREELOGIC, is available from the University of Southern California in Los Angeles.

29.4.12.1 Monte Carlo Fault Tree Analysis

It is recognized that event rates and mean down times as determinates of the likelihood of fault tree End events and thus the Top event, are *stochastic* variables. The term "stochastic" implies that the precise event rate of an End event cannot be determined as a point value. We cannot know for certain the rate that will apply over the exposure period for which the analysis is being performed. The best that can be done to estimate such variables is to place bounds on the mean value. However, it may be possible to estimate the distribution that governs the variability of End event rates and mean down times. If such a distribution can be estimated for each fault tree End event, it would be most useful if the distributions of rates and mean down times could be used as input to the tree quantification. This offers the prospect of calculating distributions of each output value of the tree. These would include Top event unavailability and unreliability or an End event Importance. Distributions of tree output values are the most complete descriptions of system performance as embodied in the tree. They allow the analyst to fix confidence bounds for Top event probabilistic measures.

The *Monte Carlo* process is most useful in modeling the performance of systems that are described by event likelihoods. For instance, if we have such a system, we may analyze its logic many times, each analysis independent, varying in a random fashion the input values and examining the number of times the output exceeds some standard. This would give us a measure of the likelihood of exceeding the standard. This fundamental process may be extended to more sophisticated methods of Monte Carlo analysis. Before proceeding with an example we should be aware of the nature of the *random* input to such a model.

Earlier statistical texts have published tables of random numbers useful in manual Monte Carlo sampling. Computers, languages, and even some hand-held calculators now contain simple algorithms that calculate a string of pseudo-random numbers. Such an algorithm need only be seeded with a starting value to calculate a string of numbers that have a uniform distribution and do not repeat over a suitably long cycle. The algorithm can be adjusted so that the numbers are bounded by 0.0 and 1.0.

A suitably indexed pseudo-random number generator will generate numbers in a uniform distribution over the interval 0.0 to 1.0 as shown in Figure 29.32. Having such an unbiased number base, we may use it to select a rate or mean down time of a particular End event from the event distribution.

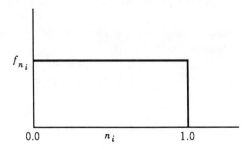

Figure 29.32 Pseudo-random numbers.

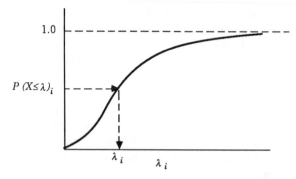

Figure 29.33 Cumulative distribution of lognormal event rate.

For instance, let us assume that a particular End event has a lognormal distribution. We wish to select rates from this distribution by the Monte Carlo process that will be combined with other rates selected in a like manner from all other End events. Using these selected End events, the tree will be analyzed.

Figure 15.1 shows a characteristic lognormal distribution as applied to an End event rate. The cumulative distribution is shown in Figure 29.33.

The Monte Carlo sampling process commences with a pseudo-random number generated with equal likelihood over the interval 0.0 to 1.0. This number is then used to select an event rate from the cumulative distribution of that particular End event. The process is shown graphically on Figure 29.33.

Performing such a selection on each End event will allow calculation through the tree for a selected output value. Repeated samples from each End event distribution for both rates and mean down times will allow a calculation of a distribution of a selected tree descriptor, such as unreliability. It would be expected that thousands of samples would be necessary to develop the output distribution. However, there are Monte Carlo sampling methods, such as Latin Hypercube, that allow a reduction to a sample size of 100 while providing an accurate distribution.

There are fewer fault tree Monte Carlo programs available than there are quantification programs with point values. There is one available for the microcomputer that is part of the Treelogic series called TREELOGIC Monte Carlo.

29.5 FAULT TREE EXAMPLE PROBLEM

An example of fault tree calculations will illustrate the methods and provide a format by which calculations are facilitated.

We will assume two minimal cut sets, one of order 2 and one of order 3. We will also assume λs and τs and an exposure time of 500 hours.

$$C_1 = AB$$

$$C_2 = BCD$$

Event	λ	τ
A	$2.3E{-}3$	.5
B	$4.1E{-}3$	1.2
C	$2.0E{-}2$	.5
D	$6.0E{-}2$	.4

The event calculations are best done in table form. This organizes your thinking and reduces errors.

Event	$\bar{a}_i$	$\bar{r}_i$
A	$1.148E{-}3$	.6834
B	$4.896E{-}3$	.8713
C	$9.901E{-}3$	.9999
D	$2.344E{-}2$	1.000

At this point, pause and note the relationships between $\bar{a}_i$, and $\bar{r}_i$. Examining rates and mean down times, it may be possible to see reasons for these values.

Calculating $\bar{a}_i$ and $\bar{r}_i$ without repair we can see they are equal to $\bar{r}_i$ with or without repair.

Next, calculating cut set values with repair and noting that $\bar{A}_k$ is integral of rate of fault whereas $\bar{R}_k$ is merely the product of $\bar{r}_i$s in the cut set, we have:

Cut Set	$\bar{A}_k$	$\bar{R}_k$
AB	$5.624E{-}6$	$7.967E{-}3$
BCD	$1.136E{-}6$	$3.030E{-}3$

Calculating cut set values without repair we have:

Cut Set	$\bar{A}_k$	$\bar{R}_k$
AB	.5934	.5934
BCD	.8712	.8712

Again, we can note that without repair there is no difference between $\bar{A}_k$ and $\bar{R}_k$.

Calculating Top event values with repair we have:

$$\overline{A}_T = 6.760E-6$$

$$\overline{R}_T = 1.099E-2$$

We can note as would be expected, that $\overline{A}_T$ is smaller than $\overline{R}_T$. Finally, calculating Top event values without repair we have:

$$\overline{A}_T = \overline{R}_T = .9479$$

Next, we should examine the Importance of End events with and without repair. However, to do this we must first find the Importance of the cut sets. Calculating the Importance of cut sets and End events with repair for both unreliability and unavailability:

Cut Set	I_k (unavailability)	I_k (unreliability)
AB	.8319	.7245
BCD	.1681	.2755

It is clear that Importance calculations are dependent on underlying values. Calculating cut set Importance without repair we have:

Cut Set	I_k
AB	.4060
BCD	.5940

Without repair, there is only one set of Importance calculations. Calculating event Importance with repair we have:

Event	I_e (unavailability)	I_e (unreliability)
A	.8319	.7245
B	1.0000	1.0000
C	.1681	.2755
D	.1681	.2755

Calculating event importance without repair we have:

Event	I_e
A	.4060
B	1.0000
C	.5940
D	.5940

30

SOFTWARE HAZARD ANALYSIS

The traditional approach to the hazard analysis of complex electromechanical systems is to treat electronic devices that process or originate system control signals as black boxes. This approach precludes analysis of the internal functioning of the box. Output reliability of the box is established relative to input and this value inserted into the system model as a quantitative representation of a pseudo-mechanical component of the system. When we replace the box with a small computer, processing instructions that have been permanently installed in the memory, we have quite another situation. If the software containing the instructions is error or fault free, then this component cannot fail and the statistical concept of measuring stochastic wearouts has no meaning. However, the software instructions may contain faults. This possibility will require an analysis of the computer program code that has become instructions to the hardware system.

For instance, an automobile engine may be controlled by a small microprocessor that will sense atmospheric density, present engine temperature, RPM, mixture, timing, and perhaps power loading and transmission gearing. In response to the driver's new power demand, this processor analyzes these input signals and provides output signals to the mixture control, spark timing, injector fuel flow, transmission, and throttle opening to match, the engine power output to the driver's demand without exceeding the engine's operating limits in any of several areas. Therefore, to analyze this power control system for the hazard of *inadvertent full power operation*, more than a simple analysis of the throttle linkage is needed.

The commands stored within small processors have their origins in computer code. This code is a series of instructions to a hypothetical computer that can process input signals from the system to output control signals. Following validation testing on a conventional computer, this code can be implanted into a chip in the form of read-only memory (ROM). This results in permanent storage of a computer program that will respond to and process signals in the operation of some part of a system. Such storage of computer-based instructions is called firmware as opposed to the software of the originally created code. However, in larger and more complex systems, such as those found in aerospace or the larger industrial systems such as electrical generating or refinery facilities, the instructions may be stored in a small, but very powerful, computer. In this case, it remains software.

Whether the coded instructions are stored in software or firmware form, analysis of the system in question for hazardous occurrences must include an analysis of the stored, coded instructions if one is to hope for an understanding of the system's potential for hazardous losses, a form of hazard analysis called *software safety analysis*.

Software or firmware are not intrinsically hazardous. They contribute to hazards through unexpected and undesired outputs. Such faulty outputs can be merely annoying, degrade efficiency, or cause losses. Hazards arise when software miscontrols a system in which it is embedded to cause an undesired event that may result in injury, property damage, or some form of environmental damage. Thus, while software may seem benign, it possesses the same potential to inflict harm as a failed set of automobile brakes.

Software possesses a unique property to be considered when analyzing it as a system component. It does not fail through wear out as a result of operating exposure. It contains *faults*. Whereas metal fatigues, electrical contacts corrode, intercoolers lose efficiency through buildup of scale, software continues to function as designed without any form of alteration unless the hardware that processes its signals fails or a magnetic field alters its statements. Thus, software faults should not be analyzed statistically. This again argues for discarding the black box approach to system analysis of units that contain software.

The purpose of software safety is to control the software faults to a level that will reduce system risk due to software malfunctions, to an acceptable level. A software fault causing a resultant harmful system function is a *software hazard*.

SOME CONSIDERATIONS OF THE PROBLEM

Software is coded by programmers working to a specification set forth by system designers. Software faults may take three forms.

1. The so-called honest errors made by the programmer in coding the software specification. These are simple mistakes in the coding process that result in the software behaving in a manner other than that which the programmer intended.
2. Faults due to incorrect software specifications or the programmer's interpretation of these specifications. These errors may result from system designer's lack of full understanding of system function or from the programmer's failure to fully comprehend the manner in which the software will be implemented

or the instructions executed. In this type of fault the software statements are written as intended by the programmer.

3. Faults due to hardware failure. Hardware failures may change software coding. Thus such software faults are secondary in that they originate outside the software.

Discovery and correction of fault types (1) and (2) is not an easy task and, in large pieces of software, may not ever be totally accomplished. Of course, fault type (3) is not present until the system is operated. A rather facetious corollary of "Murphy's law" states that: "Any software that is obviously correct, will contain faults." Although one may smile at this saying, it contains a certain perverse truth that can cause substantial harm. It is particularly true when applied to large pieces of software that have been thoroughly tested and operationally well used. In such frequently used programs, faults seem to be discovered at the moment in which they can do the most harm such as during emergency or abnormal operations when the parent system exercises a seldom used branch of the software. (Murphy says, "software hazards will always arise in a way to do the most harm.")

A software hazard may be one of the following four types:

1. An undesired signal causes an unwanted event in the system functional process.
2. An undesired signal causes an out-of-sequence event.
3. An undesired signal prevents the occurrence of a needed event.
4. An undesired signal causes an event that in magnitude or direction is out of tolerance.

Traditionally, the failure of a computer has been viewed as a hardware problem much as the failure of a television set. Although computer failures do have hardware elements, it should be recognized that a fault in the software part of a computer system must be addressed as distinct from a failure in its hardware elements. Testing and repair of computer hardware will not correct problems that originated with the software. However, testing and correction of computer hardware is necessary prior to discovery and correction of type (3) faults.

The fact that software does not fail due to internal causes does not make safety analysis of this complex subsystem easier. In theory, a properly designed testing program could discover all faults. In practice, this may be impossible due to the innumerable interactions of signals and the microsecond timing of these signals. Furthermore, as testing methods tend to be directed to the specifications to which the software was written, errors in these specifications can only be tested by full system operation in its normal operating environment. This must include emergency procedures, that are in a seldom used area of software where faults may hide until they can do the most damage.

In addition to the four generic classes of software faults set forth, there are several more direct and practical causes of software faults.

Unrealistic assumptions concerning requirements and specifications

Ignoring finite precision of a program

Failure to perform required function

Performing functions not required

Failure to properly respond to hazardous conditions in the system

Improper consideration of the system computer environment

30.1   ELEMENTS OF A SOFTWARE SAFETY PROGRAM

A software development program for a system will parallel the hardware development. Software hazard analysis tasks must be performed in each phase of the development if the system contains substantial software.

The *software requirements* phase of software development establishes the scope and tasks of the system software modules. Incorrect or inadequate specifications of software functions are the major source of software errors. These are the type (2) errors. Therefore, it is in this phase of software development that substantial progress can be made in creating safe software. Although initial requirements may be frustratingly general, they can be useful when used in conjunction with *functional flow diagrams*. Such diagrams will depict software functions within each module. Of equal importance, such diagrams can be used to depict the hardware to software, software to software, and software to hardware functional interfaces throughout the system. They can form the basis for discussions of tradeoffs of system functions between hardware and software subsystems and modules. They then become the basis for properly specifying all safety critical functions of the software modules.

A general form of software hazard analysis might be called *Software Hazardous Effects Analysis (SHEA)*. This analysis is started in the requirements phase of software development. It may depend initially on safety checklists constructed from the experience and knowledge of the system engineers. Such checklists will also provide guidance to creating appropriate software requirements. Safety checklists may contain items such as the following.

Validation of critical commands

Self-test to assure memory integrity

Generation of critical status to operator

Preclude inadvertent entry into data control

Prevent automatic control until all data is loaded

Anomaly detection that will revert to safe condition

Validity checks for operator inputs

Structured interface handshakes to ensure exchange of data and commands

Timing checks of input/output data to ensure proper functional flow

SHEA concentrates on potential safety problem areas in the software. It can be developed as part of a team effort at various levels of detail. SHEA can be performed in part by system engineers with general knowledge of software purpose and functions. However, software programmers must be involved throughout the analysis. The elements of the SHEA are the following.

Software Function. The software function within a specified module is succinctly identified.

Function Description. A brief summary of the function to include identification of any command, data input, or control function in this module.

System Hazard. Identification of the system hazard that could occur as a result of improper operation of this function.

Hazard Category. Attempt to identify the hazard category. If the hazard carries across a system interface and the hazard is not identifiable, state this in the form by words or sign.

Safety Impact. Brief discussion of the hazard to include logic for safety requirements.

Recommended Requirements. Recommended safety countermeasures to eliminate or control this hazard to an acceptable level. If control cannot be effected within software, suggest external controls.

Remarks. Any additional explanatory comments. These usually pertain to control of hazard.

Status. Indication as to whether hazard has been satisfactorily addressed.

The analysis may be documented in a conventional columnar format or in any other format that clearly depicts the information and is acceptable to management.

The first step in the analysis is to identify the safety critical areas of the system and their functional paths. These paths may contain both hardware and software elements. The analysis will focus on the software functions within each system functional flow path. The SHEA in the concept phase of system development will result in qualitative identification of hazardous potentials in the software and development of a general approach to control or eliminate them.

In the *definition* phase of system development the details of the algorithms, logic structure, and input and output parameters across interfaces of system components and software modules will be established. The software modules will be coded.

SHEA, as extended into the *development* phase, will become more specific in its identification of software hazards and their control. Fault tree analysis will be introduced at this phase. The fault tree will originate with a Top event that will be hardware based. As the system evolves and the software command, control, and monitor functions are refined, software events will begin to appear in the tree. The fault tree process will facilitate an examination of the human–software–hardware interfaces throughout the system as well as an examination of hazardous paths within each of these elements. As the software modules are coded, fault trees can be extended into the code to examine fault paths. A fault tree examination of software code for fault paths is sometimes called a *soft tree*.

The goal of a soft tree is to show that the logic planned for the software code will not cause a hazardous output. The soft tree will not locate type (1) faults. The tree will commence with a fault output from the software module and proceed down through the module modeling all possible causes of this output. Such an output may not in itself cause an accidental loss, but it may do so when it occurs in conjunction with certain human, environmental, or hardware events as discovered by the complete fault tree for the system. By locating all possible causes of each possible fault output from a software module through the Boolean logic model of the fault tree, care can be taken in the coding of the module to eliminate or tolerate these causes.

In this phase, SHEA should examine and evaluate the ability or inability of software to detect status of safety-critical devices or functions, entry into improper routines, improper sequencing or timing, and the traps or corrective actions taken to control or eliminate a system fault path.

SHEA does not address type (1) software faults.

During system development and as it reaches the *integration and testing* phases of development, software modules will be tested separately and in the system environment, debugging the code, and assuring that the software is functioning at the level of reliability and safety required. The testing should be conducted under no-fault conditions then with the likely faults that may occur in the operational environment inserted. Testing is, of course, the final check of code debugging efforts as well as the proof of all methods of hazard control. Type (1) faults should be found and corrected during this phase.

Although there are no tests that can prove the software totally dependable and fault free, the adequacy of safety-critical functions can be shown to control hazards under the range of operational conditions tested. All safety interlocks, traps, and control functions that have been designed to be fault tolerant should be extensively tested. It is important that all software module functions be tested at least once.

These software analysis tasks should be linked to a particular phase so that the software hazards may be found and controlled as early in the system development process as possible. Department of Defense systems have a carefully specified sequence of analyses linked to each phase of system development. (These are established in MIL–STD–882B, Notice 1). Although these tasks and their linked phases are specified for DOD systems, they are logical and proper in their discovery and control of software hazards. They would be most useful on complex systems.

Software Requirements Hazard Analysis (SRHA) is performed in the concept phase of system development. It may be continued, as the system is further defined, in the *definition* and *development* phases. The purpose of the analysis is to find unsafe software modes of function based upon faulty requirements. The faults found by such analysis will be based on improper specifications of the system software functions and linkages and thus be of type (2). These may be faults such as out-of-sequence, inappropriate magnitude, improper control response, improper response to environments, and other forms of command and control originated fault hazards.

SRHA will use the results of a system level PHA or any other data or analysis results that are available. The output of the analysis will be a set of hazardous effects linked to improper/inadequate specifications of the system software. In addition to analysis of system requirements as they pertain to software development, SRHA should consider the following:

1. Software requirements and specifications to software development documentation
2. All documents that describe interfaces between software and the operator or hardware
3. System functional flow diagrams
4. System safety requirements
5. Information concerning system energy sources, toxic substances, or environmental hazardous event sources that may relate to or be controlled by software

As a further output of this analysis there should be developed:

1. Safety related design and test requirements and specifications to be incorporated into the Software top-level and software detailed design hazard analysis as well as the software test plan
2. Change recommendations for system specifications that are related to software
3. Safety-related recommendations for hardware design that are related to software safety
4. Software testing requirements that are to be incorporated into system test documents

The documented form of the analysis is at the option of the performing agency. However, as a minimum it should contain the following:

1. Specification elements or system functions
2. Related software structure
3. Related software functions
4. Possible hazardous effects
5. Recommended control of hazards
6. Related test requirements

Development of items (1) to (4) were not adequately covered by the five elements of the primary analysis.

SRHA should be complete before the remaining forms of analysis are initiated.

Top-Level Design Hazard Analysis (TDHA) will examine the design of each software functional module for hazards. It will use as input the results of the SRHA as well as other documentation used for the SRHA. This analysis will be primarily performed during *definition* phase of the system life cycle. It may be continued during the *development* phase.

The analysis will define software modules and examine their functions for possible hazardous effects. In this definition process, possible implementation of each software functional module into a code design will be considered. Any software module that can influence system function to a hazardous outcome will be considered safety critical.

Safety-critical software modules will be further analyzed for control of the related hazards. All hazards previously developed in the PHA, SSHA, or SRHA will be examined for their relationship to a particular software module or sets of modules. Independence or dependence of these modules will be investigated as part of the TDHA. The top-level design of each software module will be analyzed for creation of hazardous possibilities that may be propagated down through the system to hardware and a resultant accident. The nature or functional flow of the top-level design of computer modules should be examined for the generation of unacceptable hazards. All forms of hazard generation delineated above such as timing, or out-of-sequence events should be considered in this examination.

The documented form of the analysis is at the option of the performing agency. However, as a minimum it should contain the following:

1. Software functional modules
2. Possible hazardous effects
3. Software module dependencies
4. Possible software design methods for control of hazards
5. Recommended test requirements

TDHA should be completed prior to the initiation of lower forms of software hazard analysis.

Detailed Design Hazard Analysis (DDHA) shall be performed following the TDHA to examine the implementation of methods of developing acceptably safe software set forth in the TDHA. This analysis will be performed during the *definition* and *development* phases of the system life cycle. The analysis will examine the software modules specified in the TDHA to determine if, in the implementation of their functions, system hazards will remain within acceptable limits and overall system risk due to software faults is acceptable. This analysis does not examine code, nor should code be available for examination prior to the completion of the DDHA. Its purpose is to fix the nature of the coding so that acceptable risk of system operation may be achieved.

The analysis will use as inputs the results of the PHA, SSHA, SHA, SRHA, and TDHA as well as other system specifications and documentation. Hazardous possibilities developed and examined during the TDHA will be followed through development of functional implementation in possible coding schemes to assure that they are adequately controlled by the coding methods selected. This analysis is the final opportunity to fix the safety requirements for the coding of the software functions that will follow. As in prior analyses, potential software faults due to timing, out of sequence, or other of a number of causes of software hazards delineated above will be examined for control of their hazardous possibilities during the coding.

The documented form of the analysis is at the option of the performing agency. However, as a minimum it should contain the following:

1. Software modules
2. Possible hazardous effects
3. Recommended coding approaches to control hazards
4. Identify safety-critical software modules
4. Recommended test requirements

DDHA should be complete prior to initiating the lower forms of software hazard analysis. The results of this analysis should be used to develop safety-related information for inclusion in user manuals. These would include operator manuals, diagnostic or maintenance manuals, and other types of computer support manuals. Specific, safety-related coding recommendations should also evolve from this form of analysis.

Code-Level Software Hazard Analysis (CSHA) shall be performed following the DDHA to assure that the coding of each software module is completed without introduction of unacceptable hazards. This analysis will be performed primarily during the *development* phase of system life cycle. The analysis will analyze program code for all software modules to assure that requirements of the previous analyses have been met in the execution of the coding and that no further hazards are introduced in the coding process. The analysis will consider the outputs of the TDHA and DDHA analyses as establishing the nature of hazards that must be controlled to an acceptable level during the coding process. The analysis will attempt to assure that no hazards of type (1) are introduced during coding. This will be done with code walk-throughs and other methods of code examination for desired functional output.

Specifically, the CSHA will:

Analyze computer modules for timing, out of sequence, or other improper outputs from the software modules.

Ensure software implementation of safety criteria called out in the previous analyses.

Ensure the required degree of fault tolerance being incorporated into the code.

Ensure that the code is as desired by the programmer.

This will be the last opportunity to examine and correct the code prior to system testing. Therefore, it is important to assure that the code does not degrade the safety of the system below that which was deemed acceptable in the earlier analyses.

30.2 PETRI NETS

A *Petri net* is a mathematical model of a system developed by Carl Petri. The modeler describes the system in graphical symbols and need not be concerned with the mathematical foundation of the system. A Petri net has a number of advantages:

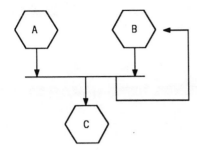

Figure 30.1   Petri net example.

1. It can model system hardware, software, and human components.
2. It can be used early in the development cycle. This facilitates changes in system design.
3. It may be quantified with probabilities and timing.
4. It is effective on synchronization problems in system modules, as well as fault tolerance.

A Petri net is modeled in terms of conditions and events. If certain conditions are in effect, then an event or state transition will take place. After the transition other conditions may well be in effect. The net is modeled with a few graphical symbols. A *place* in Petri net terminology is a condition and is modeled by a circle. If a place contains a *token*, it can be considered to be the current state of the system. An event is a *transition* represented by a bar. An *arrow* from a place to a transition indicates that the condition is an input to the transition. An arrow from the transition to a place indicates posttransition state of the system or a postcondition (see Figure 30.1).

Figure 30.1 states that if places A and B are true, a transition can take place. After the transition places B and C are true. A token or "∗" can be used in the net to indicate that a place is currently true. If a place holds it will be shown by a token. Figure 30.2 illustrates.

Figure 30.2 shows that places P_1 and P_2 are needed for transition t_1, and places P_2 and P_3 are needed for transition t_2. The tokens in P_2 and P_3 show that these

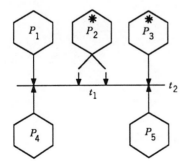

Figure 30.2   Petri net with tokens.

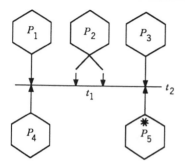

Figure 30.3 Posttransition state of Figure 30.3.

places are true. Thus, only transition t_2 can take place and the only post condition will be P_5. Figure 30.3 illustrates.

If a transition has no preconditions, it can occur at will. A Petri net can be exercised by moving the tokens through the net. At any instant, the state of the net is represented by those places that contain tokens. Fundamentally, the Petri net represents the ordering of signals that may pass through a system. This method of analysis is most suitable for analyzing a software module for fault *free* function.

For instance, using a Petri net to analyze a software module coded to assure that one event would always precede another, an analysis of the code should result in Figure 30.4.

30.3 RULES AND GUIDELINES

There are a number of rules and guidelines that can be formulated for software design and documentation to assure acceptable safety. This list must be tailored to the requirements and peculiarities of the function and environment of the software modules of a given system. This list is taken in part from the *U.S. Air Force Software System Safety Handbook*, SSH 1-1. It is quite general, yet comprehensive, covering multiple software–CPU functions.

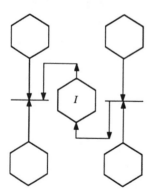

Figure 30.4 Petri event precedence.

30.3.1 Documentation of Designs and Intents

1. Safing requirements should be detailed in the software design specification.
2. Make full use of the self-test and fault-tolerant concepts.
3. Interrupt priorities and responses shall be documented and analyzed for safety impact.
4. Fully state the functional intent of the software design.
5. Critical software functions must not be corrected by patching in a language lower than the source code designated for that software. All patches must be clearly documented for meaning.
6. Any assumptions made concerning unknown or previously unstated requirements for hardware/software or their interfaces must be clearly stated.
7. Any deviations from stated requirements for hardware/software or their interfaces must be clearly stated for software module including implications and limitations.
8. Operating system software, particularly shared software such as library routines, should be completely described.

30.3.2 Software–CPU Design Rules

1. A single CPU controlling a function that could result in system loss or loss of human life should be incapable of satisfying all requirements for initiating that function. In a multiple CPU environment, two or more CPUs shall be required to initiate such a function. In a single CPU environment, other hardware or logic devices should require an intersection with the CPU commands to initiate the critical function.
2. Critical data transmitted between CPUs should successfully pass data verification checks in both CPUs.
3. Critical software code should be structured to enhance comprehension of decision logic.
4. Software should be as modular as possible to improve logic error detection and correction.
5. Memory locations should have one unique name to prevent memory usage conflicts.
6. Event sequences that tax human response times shall be controlled with software.
7. Safety-critical timing in software logic should not be changeable by an operator.
8. Upon detection of safety-critical anomaly, the system should revert to a safe configuration.
9. Upon detection of a safety-critical anomaly, the software should inform the operator of the anomaly. Noncritical anomalies shall be recorded to aid in system analysis. If the software took action as a result of a safety-critical

anomaly, the operator should be informed of the action taken and this action recorded for system analysis.

10. If the software saves the system, the resulting system configuration should be reported to the operator.

11. Saving scenarios for safety-critical hardware should be incorporated into software logic.

12. Software-controlled functional sequences affecting safety should require a minimum of two independent procedures for initiation.

13. Any software module should be initially set to a known and safe state. The software function should also terminate to a known and safe state. In anticipation of system restart or random power inputs, all devices should be set predictably safe.

14. All unused memory locations should be initialized to a pattern that if executed as an instruction will cause the system to revert to a known safe state.

15. Provisions should be made to protect the accessibility of memory regions containing data or instructions that could affect the safety of the system.

16. All flags should be unique and of single purpose.

17. Simple instructional sets for CPUs are preferred. CPUs should load entire instructions and data words rather than loading in parts.

18. CPUs should be so designed that they segregate instructions and data memories in separate banks served by different buses rather than the same memory bus to access both instructions and data from the same address space.

19. Critical operational software instructions should be resident in nonvolatile memory (ROM).

20. Data used by safety-critical software should be protected by error checking and correcting memory controller.

30.3.3 CPU–Hardware Interface Rules

1. A status check of safety-critical, one-shot functions shall be made prior to executing a potentially hazardous sequence of functions.

2. Upon completion of a test where safety-interlock functions are removed, the restoration of those functions shall be verified.

3. When software generates a hardware command, analysis shall determine if the command should be continuous until reversed or only applied for discrete time.

4. Decision logic using registers that obtain values from end item hardware or software, shall not be based on values of all "ones." Neither shall such logic be based on values that are all "zeros."

5. Separate arm and fire commands shall be required for ordinance initiation.

6. Safety-critical hardware controlled by software should be initialized to a known safe state.

7. Safety-critical input/output ports should have addresses sufficiently different from noncritical ports that a single or multiple address bit failure will not allow access to critical ports.

8. The system shall safe-state recovery from inadvertent instruction jumps.

9. The CPUs shall be immune from the effects of temporary power outages.

10. The CPUs shall be protected against electromagnetic interference.

11. There should be periodic memory integrity checks of both program and data memories if served by different buses.

12. The integrity of instruction ROM should be enhanced with a memory controller or type of check-summing algorithm.

13. There should be periodic data and program memory bus operational checks.

30.3.4   Operating System Software Rules

1. The software should discriminate between false and valid interrupts.

2. Care should be taken to prevent software working on a fixed stack size from overflowing and overwriting other program instructions.

3. Software shall place all sharable CPU resources on a stack or other prescribed memory area before a called module begins to execute with these resources.

4. Sharable software utilities such as trigonometric functions should not overwrite their temporary scratch memory areas when they are used in several levels.

5. Sharable software utilities should not internally call another software utility.

30.3.5   Operator Interface Rules

1. Software should require two or more operator responses for initiation of any potentially hazardous sequence.

2. Operator interactions with software should be as concise as possible.

3. Software should provide for detection of improper sequence requests by the operator.

4. The software should provide for operator cancellation of software processing. The cancellation should require only a single keystroke from operator.

5. Processing cancellation by operator shall always place the system in a safety state.

31

SNEAK CIRCUIT ANALYSIS

Complex electromechanical systems may contain latent functional paths that will result in unplanned and undesired modes of operation. Such functional paths are called *sneak circuits* and the process by which they are discovered is called *sneak circuit analysis* (SCA). These paths are not due to component failures. They are designed-in, the design process performed in conformance with system specifications. They do not depend on human or environmental events to generate the sneak path, although human or environmental events may activate the paths. They are usually latent in that they cause problems when seldom used circuit paths are activated. This may occur during emergency operations or system functions that are attempting to bypass a malfunction. They originate in one of the following problem areas.

Sneak Timing. Sneak timing results from simultaneous or out-of-sequence functioning of components or command signals. Frequently computer generated signals to subsystem areas will be sneak timing problems.

Sneak Path. Latent paths in circuitry that may occur without component failures. Such paths may inhibit desired functions or permit unwanted functions. Such paths frequently occur due to unanticipated ground switching operations or due to improper power supply linkages.

Sneak Indicators. An ambiguous or false status indication resulting from improper paths in status lights or meters.

Sneak Procedures. These consist of ambiguous wording, incomplete instructions, lack of warnings or cautionary notes that may lead to improper operator actions under emergency or contingency conditions.

Sneak Labels. Improper labels or imprecise nomenclature or instructions on consoles can lead to operator errors that can be compounded by inappropriate system path activation.

Sneak circuits usually originate from activities in the design and development of a system that are somewhat out of the ordinary. The following are examples of the kinds of activities that may generate sneak circuits.

Changes. Modifications to the original design after it has been once completed can result in such problems. Revisions are sometimes done without complete system integration or testing of all operational modes.

Incompatible Design. Subsystem designs prepared by different design teams or by different contractors may contain incompatibilities that surface when least expected or when they can do great harm.

Field Fixes. Malfunctions that occur during testing are sometimes corrected by field fixes. These may solve the immediate problem, but a lack of full system integration may allow new problems to be inserted into the system.

Human Errors. When human operators are in a control loop, there are possibilities for unanticipated modes of system operation. For instance, manual operations may be performed out of sequence or proper procedures may not be followed in areas that create unanticipated system functional flows.

31.1   SNEAK ANALYSIS TECHNIQUES

Sneak circuits have distinct characteristics. These characteristics enable the experienced analyst to detect and correct such sneaks by employing two fundamental approaches.

1. Recognition of basic topological patterns that may be used to describe all electrical circuitry either singly or in combination.
2. Examine each topological pattern for certain faults that could indicate the presence of sneaks.

31.1.1   Topological Patterns

It is believed that all circuitry may be resolved logically into one of five patterns or combinations of these patterns. These patterns can be examined for flow structure at each node that will indicate the presence of sneaks. The topographs assume that unswitchable power is at the top of the node and that an unswitchable ground is at the bottom. These five patterns are the straight line, ground dome, power dome, combination dome, and reverse current. These five patterns are shown in Figure 31.1.

Given each topograph of Figure 31.1, there are a set of clues or questions that may be asked in an attempt to find the sneaks that exist there. These clues are typically proprietary and there is no finite, exhaustive set. As designers and analysts gain experience with sneak analysis their clue set will be expanded.

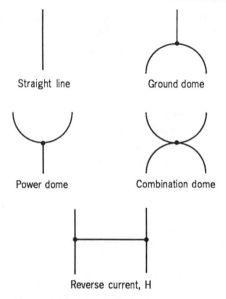

Figure 31.1   Electrical circuitry topographs.

As an example of clues for the straight-line topograph, we should examine Figure 31.2.

Note:

S_1 is a general switch such as circuit breaker or fuse.

L_1 is a general load such as logic gate output or relay coil.

A few possible clues associated with the topograph of Figure 31.2 would be the following.

Figure 31.2   Straight-line topograph with switches and loads.

1. Do power and ground originate at the same source?
2. Is switch S_1 open when load L_1 is desired?
3. Is switch S_1 closed when load L_1 is not desired?
4. Is switch S_1 necessary?
5. Does label of switch S_1 match function of load L_1?

The literature will contain more complete clue sets, however, each organization must build up its own set.

31.2   SNEAK CIRCUIT GUIDELINES

There are a few guiding principles for sneak circuit analysis. A short list of such principles is set forth as follows.

1. Refer to other system safety analyses to identify adverse End events that can result from sneak circuits. Review equipment design to identify those circuits to be analyzed.
2. Mark the partition points where different subsystems and "black boxes" interface so that system or subsystem analysis can be divided into manageable portions for detailed analyses.
3. Review the drawing of each small subsystem. By use of standard electrical symbology, prepare the data from the "as built" drawings for computer analysis.
4. Uniquely identify and encode all wire segments for computer storage. This information should include every single path segment using from–to iden- tification.
5. Process the data in the computer using an appropriate SCA program. Such programs are commercially available. The output of the computer analysis should define all possible paths that can exist in the circuitry.
6. By use of topological symbols prepare individual network trees for all circuitry. Note that these trees use a different representation from standardized electrical drawings and show such factors as circuits that can receive power; power and ground points; junctions such as loads, diodes, switches, and disconnects between power; and ground points. Network trees can be drawn automatically by a computer-controlled plotter or manually on a graphics terminal.
7. Prepare a network topological "forest" that will be a representation of inputs required to cause each undesired End event and the circuitry involved in the generation, control, and transmittal of those inputs. Let the diagram show all interfaces between units.
8. Apply sneak clues to identify and analyze sneak circuit paths.

EXERCISES FOR PART IV

1. How do you convince management that an acceptable state of safety allows for some accidental losses?

2. How should a hazard be described? Discuss the properties of each descriptor of a hazard.

3. Describe the properties of the five types of fault.

4. Consider the system and subsystem hazard analyses.
   (a) During what life cycle phases are they performed?
   (b) List the methods of analysis that can support each.
   (c) Set down an example format for each showing the kinds of information needed for each analysis.

5. What is the relationship of the preliminary hazard analysis to other forms of analysis?

6. How and where does the operating and support hazard analysis fit into the other types and methods of analysis?

7. Describe the relationship between the Boolean equivalent tree and the original fault tree.

8. Given the below Boolean expression for a fault tree top event, reduce it to an expression of the minimal cut sets.
   (a) $T = ACDE + ACDF + BCD + CD + AE + AC + ACF$
   (b) $T = ABCDEF + AB(BCD + EF + FGHJ) + BFGH(BFG + HIJ)$
   (c) $T = A(B + C + BDF)(ABD + CEFGH)$

9. Given the following minimal cut sets, find the path sets.
   (a) $A, BD, BDE, BEFG$
   (b) $ABD, BEF, BFGH$
   (c) $A, CD, CE, EFGH$

10. Define the following top event descriptors.
    (a) Unavailability
    (b) Unreliability
    (c) Expected number

11. Define the following end event descriptors.
    (a) Fault rate
    (b) Mean down time
    (c) Unavailability
    (d) Unreliability
    (e) Importance by unavailability and unreliability

12. Given the following cut sets and their rates and mean down times, compute the following:

$$C_1 = AB$$
$$C_2 = ACD$$
$$C_3 = BDE$$

Events	λ/hr	τ(hr)
A	$2.3E-3$	.5
B	$5.4E-4$	1.1
C	$3.6E-3$	.7
D	$5.4E-2$	.3
E	$4.5E-3$	1.2

Assume 145 hours of exposure time.

(a) Event, cut set and top event unavailability, assuming repair.

(b) Event and top event unreliability, assuming no repair.

(c) Event and top event unreliability, assuming repair.

(d) Compute event importance for both unavailability and unreliability with repair.

REFERENCES AND BIBLIOGRAPHY

Aggarwal, K.K., "Comments on the Analysis of Fault Trees," *IEEE Trans. Reliab. R-25*, 126–127, 1976.

AMCP-706-196, *Engineering Design Handbook, Design for Reliability*, Department of the Army, January, 1976.

Anderson, T. and Lee, P.A. *Fault Tolerance: Principles and Practice*. Prentice-Hall, Inc. Englewood Cliffs, NJ, 1981.

Apostolakis, G. and Lee, Y.T., "Methods of the Estimation of Confidence Bounds for the Top Event Unavailability of Fault Trees," *Nucl. Eng. Design 41*, 411–419, 1977.

Avizienis, A., "The N-version Approach to Fault-Tolerant Software," *IEEE Trans. Software Eng.*, SE-11, *12*, 1491–1501, December 1985.

Barbour, G.L., "Failure Modes and Effects Analysis by Matrix Methods," *Proceedings, Annual Reliability and Maintainability Symposium*, 1977.

Buratto, D.L. and Goody, S.G., "Sneak Analysis Application Guidelines," RADC-TR-82-179, June 1982.

Burdick, G.R., Marshall, N.H., and Wilson, J.R., *COMCAN-A Computer Program for Common Cause Analysis*, AeroJet Nuclear Company, 1976.

Dowd, C.R., "System Safety Engineering in Software Development," Boeing FSCM 81205, Seattle, WA, 1984.

Dussault, H.B., "The Evolution and Practical Applications of Failure Modes and Effects Analyses," RADC-TR-83-72, 1983.

Fussell, J.B., "A Formal Methodology for Fault Tree Construction," *Nucl. Sci. Eng. 52*, 421–432, 1973.

Fussell, J.B., "Fault Tree Analysis—The Secondary Failure," *Reliability and Fault Tree Analysis*, SIAM, 1975.

Fussell, J.B., "How to Hand Calculate System Reliability and Safety Characteristics," *IEEE Trans. Reliab. R-24*(3), pp. 169–174, 1975.

Henley, E.J. and Kumamoto, H. *Reliability Engineering and Risk Assessment*. Prentice-Hall, Inc., Englewood Cliffs, NJ, 1981.

Haasl, D.F., "Advanced Concepts in Fault Tree Analysis," System Safety Symposium, Seattle, WA, 1965.

Hall, F.M., Paul, R.A., and Snow, W.E., "Hardware/Software FMECA," *Proceedings, Annual Reliability and Maintainability Symposium*, p. 320, 1983.

"IEEE Standard Glossary of Software Engineering Terminology," IEEE Std. 729–1983, The Institute of Electrical and Electronic Engineers, Inc., February 18, 1983.

IEEE Std. 829–1983, "IEEE Standard for Software Test Documentation," The Institute of Electrical and Electronc Engineers, February 18, 1983.

Ireson, W.G., and Coombs, C.F. Jr., *Handbook of Reliability Engineering and Management*, McGraw Hill, New York, 1988.

Lambert, H.E., "Measures of Importance of Events and Cut Sets in Fault Trees," *Reliability and Fault Tree Analysis*, SIAM, 77–101, 1975.

Lapp, S.A., and Powers, G.J., "Computer-Aided Synthesis of Fault Trees," IEEE Trans. Reliab. R-26, 1977.

Leveson, N.G. and Stolzy, J.L., "Safety Analysis Using Petri Nets," *IEEE Trans. Software Eng.*, 1986.

Leveson, N.G., Shimeall, T.J., Stolzy, J.L., and Thomas, J. "Design for Safety Software," *Proceedings of American Institute for Astronautics and Aeronautics*, Space Sciences Meeting, Reno, NV, 1983.

MIL–STD–1629A, "Procedures for Performing a Failure Mode, Effects, and Criticality Analysis," Department of Defense, November, 1980.

Parnas, D., "Software Aspects of Strategic Defense Systems," *Commun. ACM 28*(12), 1326–1335, 1985.

Pizzo, J.T., and Moriarty, B.M., Quantification of Subjective Judgements for Safety Estimates-Use of Metric Matrix Hierachrical Analysis (MMHA) Proceedings of Seventh International System Safety Conference, San Jose, CA, 1985.

Roberts, N.H., Haasl, D.F., Vesely, W.E., and Goldberg, F.F., *Fault Tree Handbook*. U.S. Nuclear Regulatory Commission, Washington, DC, 1980.

Roland, H.E. and Philipson, L.L., Investigation of the Feasibility of Application of the Digraph-Fault Tree Synthesis Process to Transportation Accident Occurrence Modeling," NHTSA, 1980.

Roland, H.E. "Documentation for Computer Programs for Fault Tree Analysis with Accompanying Programs," NHTSA, 1980.

Salem, S.L., Apostolakis, G.E., and Okrent, D., "A New Methodology for the Computer-Aided Construction of Fault Trees," *Annals of Nuclear Energy, 4*, 417–433, 1977.

U.S. Air Force Inspection and Safety Center, "Software System Safety, AFISC SSH 1-1, 1985.

Wagner, D.P., Cate, C.L., and Fussell, J.B., Common Cause Failure Analysis for Complex Systems," *Nuclear Systems Reliability Engineering and Risk Assessment*, SIAM, p. 289, 1977.

V

RISK ANALYSIS

32

RISK ASSESSMENT
IN SAFETY

32.1   INTRODUCTION

In the not-too-distant past, many have accepted undesired events as inevitable or predestined. In the recent past, we have thought it too difficult to predict future losses and even if predicted, nearly impossible to mitigate. We can now appreciate that these, almost traditional, approaches to loss are inadequate. Losses may not only be predicted, but their magnitudes reduced to an acceptable level.

Our purpose in adding a risk section to this book is twofold. First, we believe that a book on system safety would be sadly lacking if it did not address the subject. In this section, we attempt to impart to the reader how risk complements system safety, yet how it differs. We wish the reader to understand that risk has a somewhat larger purview than does system safety.

Second, there are some subjects in system safety that have not been discussed, but that we must address in the context of risk. For instance, one may note that measures of safety performance developed in the foregoing are largely likelihood measures of event occurrence. The questions of the consequence or the loss due to an undesired event occurrence have been addressed too briefly. We have not considered the multidimensional nature of measuring safety performance and making decisions in this arena, nor have we considered the value of time in making decisions concerning the future behavior of complex systems.

32.2   NATURE OF RISK

The term "risk" has become in recent times nearly a buzzword. It is applied to such diverse situations that the lay person becomes confused as to its proper meaning. There is the risk of buying a new home. There is the risk of starting a new business.

There is the risk of cancer, the risk of flying. One may find oneself thinking of the risk of an expert statement being untrue. In system safety we have the risk of operating a system. Obviously we should begin this discussion of risk with an examination of the proper definitions of risk.

Risk has a connotation of undesired outcome. From this general description of the term there emerges only two proper definitions of risk. The other applications are derived from these two meanings. One proper meaning is the measure of undesired uncertainty in measuring a situation or estimating a parameter. Thus one may speak of the risk that a certain accident rate is greater than the true rate. One would hope that such a risk is a small number, say probability .05. In such a case a safety professional could use the given or computed rate knowing that there was only a probability .05 that the situation was worse than planned. If we speak of the risk of achieving a specified rate of return on an investment we should properly be speaking of the probability of a return less than the specified value.

The other proper definition of the term risk is to describe a potential loss that may occur under uncertain circumstances in the future. Such a use of risk has two elements—likelihood and degree of harm. It would be improper to speak of the risk of motor vehicle death last year. That matter is known and thus there is no element of risk in such a statement. If we are to speak of the risk of motor vehicle death next year it should be couched in a special term that combines both likelihood and harm.

An expected value is the product of the likelihood and degree of harm of a future event. More precisely, an expected value can be calculated by finding the product of the probability of an event and the value of the event. If the event has a loss that can be estimated, then the probability of that event multiplied by its loss is the expected value of the occurrence of that event. In the case of motor vehicle deaths next year, we may say that the expected deaths by motor vehicles next year is 45,000.

We may think of this number as a mean value of a stochastic variable. Thus, if the distribution of this variable is not too severely skewed, there is approximately a .5 probability that the number of deaths next year will be greater than 45,000 and a probability .5 that the number of deaths will be less than 45,000.

If one wishes to calculate the risk of death in operating a given system, it would be necessary to calculate the probability of an accident causing death with this system and multiply this number by the the number of deaths in each accident. Operating more systems over greater exposure periods would increase the probability value used in this calculation and thus increase the expected numbers of death or the risk of death.

The total risk in system operation would also include injury and property damage. Such a calculation will require that death, injury, and property damage be converted to a unidimensional measure of value. We would normally use dollars, although some calculations may find it convenient to use measures of utility value. Such a calculation would involve a summing of the expected values of loss for death, injury, and property damage of system operation. The probability values in such calculations must be based on a given exposure period. Thus, if we wished to calculate the expected value or risk of system operation in a year, we would use probability values based on a year's exposure of these systems to the potential of loss.

This form of risk calculation may find us calculating a risk of death from a certain event within a year as a finite value when we have yet to experience any deaths from this event. For instance, analysis may show us that the probability of a major meltdown in a nuclear power plant is $E-4$ per year in the United States. If we expect 100,000 deaths from this event, then the expected number of deaths in any year from this type of event is 10. This value can properly be called the risk of death per year from this event and it is appreciable even though we have experienced no deaths in the United States.

Most other uses of the term "risk" involve the expected value of potential loss. For instance, insurance risk is a term that may be applied to the expected value of loss assumed by an insurance policy or transferred by this policy. The risk of a financial investment is the expected value of loss that may occur if the investment goes bad.

Rational people are, by nature, risk averse. Some gamblers are not but we will not attempt to address this aberrant form of behavior. While people seek to avoid risk they also seek gain. By *gain* we mean an outcome of a decision or the operation of a system that produces a desired or positive result. One can say that this outcome has positive utility value. It would be extremely rare to have gain without risk. If such was the case so many would participate in this activity and the gain would be reduced toward zero. The operation of all complex systems involves risk. Of course, they would not be operated without an accompanying gain. An acceptable state of safety is one in which the gain is sufficient to make the operation of the system desirable even though there is appreciable risk. Thus, we operate passenger aircraft, although each year one or more accidents kill hundreds of people.

All individuals, businesses, and governments function in a balance between risk avoidance and gain from risk acceptance. Although a very wise person or organization may thwart this balancing process by making wise decisions, such wisdom is unusual. In the normal course of events, unusual gain accrues from accepting unusual risks. If the system generating gain or benefits is operating against nature, we may suppose that nature is indifferent to the generation of risk. However, if the benefits are significant the risk will not be insignificant or others will participate making the benefits less.

If the focus is entirely on risk avoidance with no consideration of gain, no risk is acceptable. If gain is valued too highly, then risk is taken that should not be accepted. If such a risky course is taken repeatedly, the inevitable result will be a disastrous net loss. Thus, we find that an optimal balance between risk and gain is essential to the continued functioning of an individual, system, or business. A lack of recognition that this balancing process—with its accompanying payoffs, negative or positive—is always ongoing, can place individuals or organizations in jeopardy.

However, before we can perform such a balancing process we must have risk evaluation to a degree of accuracy needed to meet the requirements of the decision process.

32.3   RISK ASSESSMENT MODEL

The concept that risk represents a potential for loss in the uncertain future is logically simple. However, the assessment of risk is logically quite complex. This is due to a number of factors, not the least of which is the complex nature of the accident process and its prediction. To assist in an understanding of this assessment process, we will set forth the elements of the risk function in somewhat greater detail than will be used to make calculations.

A conceptual model of risk assessment should contain no less than the elements shown in Figure 32.1

Fault events are the low-order primary events that, when they occur in certain combinations, cause the initiating event. Fault events may be the End events of a fault tree.

Initiating events correspond to the Top events of a fault tree. The initiating event may be thought of as an event that deviates from normal in the operation of a system. It is an unplanned and undesired event. It has the potential to cause death, injury, or property damage. An initiating event may be the engine failure of an aircraft. The fault events of such an initiating event are all the possible combinations of events (cut sets) that may cause such an initiating event. The fault events then are the human, electromechanical, and environmental fault events that in set combinations could cause the engine failure. Of course, an evaluation of the risk of operating an aircraft would require many initiating events beyond that of engine failure. Each of these would have their own set of fault events.

A risk assessment of a large and complex system is usually limited to consideration of only the most potentially damaging of initiating events. The total risk of operating an aircraft will contain many less damaging initiating events. These would involve subsystem failures that would prevent the aircraft from performing its mission but cause no human or property damage except for the necessary repairs of the subsystem. The majority of the risks will usually fall into the more damaging of the initiating events.

Consequences are the immediate results of the occurrence of the initiating event. Consequences for the same initiating event vary. When an aircraft engine fails, it may cause a crash with large consequences or the aircraft may be landed successfully with little damage. Failure of brakes on a motor vehicle may result in nothing more than a rolling stop. It can result in a devastating crash.

Losses describe the results of consequences. An aircraft crash as the result of an engine failure can have a variety of losses. They encompass the usual, death, injury, or property damage. The crash may occur in an open field or into an inhabited area causing further variation in the losses. Losses will include damage to the environment.

Figure 32.1   Risk assessment model.

Costs are the values that are placed on the magnitudes of the different classes of losses. Assessment of costs raises difficult questions. What is the value of a human life? Should the value be a variable depending on the circumstances of the individual that has died and the circumstances of his or her dying? One point of view as to the value of life in the United States is to take the Gross National Product of the United States and divide by the population. Such a calculation will give a small value of life. If we take the mean of court awards from civil litigation resulting from death, we will have a somewhat larger value. Generally, the courts will judge that each life has its own unique value.

It is easier to assess the value of injury and property damage. Property damage poses a dilemma in that we must determine whether property should be valued at replacement or acquisition cost. Do we depreciate older equipment? There is the question of how to value hidden, indirect costs of property damage. An aircraft lost from a commercial airline usually cannot be replaced with a purchase or lease cost equal to that of the lost aircraft. Such indirect costs apply to the loss of a valuable worker in an accident. The replacement will be at a cost. However the problem is approached, we must place unidimensional measures of value on losses if we are to evaluate risk.

A solution may be found in utility value theory. This body of knowledge describes how value can be placed on any item, from positive to negative. The units of value are called utiles and may be scaled to a suitable size.

A conceptual or theoretical model of the risk evaluation process should take the following form.

$$P(Ct_n) = \overset{i\,j\,k}{\Sigma\Sigma\Sigma} P(I_i)P(C_j/I_i)P(L_k/C_j)P(Ct_n/L_k) \qquad (32.1)$$

where

n $\quad\quad\;\;$ = loss type
$P(Ct_n)$ $\quad$ = probability of cost, Ct_n per exposure unit
$P(I_i)$ $\quad\quad$ = probability in initiating event, I_i
$P(C_j/I_i)$ $\;\;$ = conditional probability of consequence, C_j, given I_i
$P(L_k/C_j)$ $\;$ = conditional probability of loss, L_k, given C_j
$P(Ct_n/L_k)$ = conditional probability of cost, Ct_n, given L_k

The expected loss per exposure unit may be computed by:

$$E(Ct_e) = \overset{n}{\Sigma} P(Ct_n)Ct_n \qquad (32.2)$$

Total risk may be determined by evaluating the expected cost of the loss, $E(Ct_e)$ over all exposure units.

$$Risk = \overset{e}{\Sigma} E(Ct_e) \qquad (32.3)$$

where e = exposure units.

If it was possible to evaluate equation (32.3), there would be a precise evaluation of risk. This model could be modified to calculate risk into its three dimensions deaths, injuries, and property and environmental values. This latter form of risk modeling is most common, although the above precise form of conditional probabilistic modeling is seldom used.

Equations (32.1–32.3) do not explicitly recognize the stochastic nature of the variables that make up the idealized risk model. Such a model would replace summations with integrals.

The foregoing model takes care to recognize exposure as the key parameter in generating the opportunity for accidental events and thus risk. An aircraft in a hangar represents no opportunity for the system to generate a loss through operation. Neglecting the residual hazards of the hangar such as fire or electrical problems, such a system possesses no potential for loss. The greater the exposure the greater the loss potential and thus the risk.

The term risk as used by the insurance industry has the same basic dimensions as the above risk model. If one were to carefully consider the risk assumed by an insurance policy for personal or industrial risk, the same elements would be found lurking behind the establishment of the premium for this risk transfer to the insurance company. These would be a consideration of the initiating events that would cause a loss, the possible consequences of these events, the losses resulting from these consequences, and finally the cost of these losses to the insurance company. Furthermore, the premium would undoubtedly be set considering the exposure of the individual or industry. More workers covered in a given industry would result in higher premium. More miles driven on an automobile will, in some insurance companies, result in higher premiums. Although an insurance company will resort to standardized tables for setting premiums in many situations, the foregoing factors must be directly or indirectly considered in setting the amount of risk to be transferred to the company and thus the premium paid by the insured for the risk transferred.

32.4   RISK DECISION PROCESS

Before a decision can be made concerning system risks, the risks must be evaluated. A major element of the evaluation process is perception of risk. It may be said that:

If the public or interest group perceives a risk to be unacceptable, it is unacceptable.

Perceived risks are very real to the perceivers. Certain individuals perceive the risk of flying to be unacceptable, although they can be shown that the risk level is quite low and that their fear is irrational given that they accept the risk of driving. There are many factors that affect our perceptions of risk and our tendency to find them more or less acceptable. For instance, people tend to find voluntary risks more acceptable than involuntary; natural risks more so than man-made; delayed effect more so than immediate; on the job more so than off the job.

Table 32.1. Risk perception factors

More Acceptable	Less Acceptable
Voluntary	Involuntary
Natural	Man-made
Controllable	Uncontrollable
Delayed effect	Immediate effect
You and yours	Me and mine
Essential	Nonessential
Off the job	On the job
Sensory perceived	Sensory unperceived
Relates to self-worth	Does not relate
Greater benefits	Lesser benefits
Experienced	Not experienced
Understood	Not understood
No alternatives	Alternatives
Underestimate	Overestimate
Understood	Not understood
Known	Unknown
Common	Uncommon
Mundane	Dramatic
Little media coverage	Much media coverage
Noncontroversial	Controversial
Me in control	You in control
Voluntary	Involuntary
Fun	Work
Benign experience	Hurtful experience

A related factor of perception is under or overestimation of the magnitude of a risk. We tend to underestimate the level of risk for systems with which we are well acquainted, overestimate for the unknown; underestimate those that can be controlled or are fun; overestimate those that are not understood or are dreary.

Listed in Table 32.1 are sets of risk perception factors that will allow us to adjust our judgment when evaluating and assessing risks.

We have previously stated that risk acceptance is based, in part, upon gain. However, certain activities have a requirement or a known or well-established gain. In these cases, the issue of risk acceptance rests upon the nature of the risk because the gain is a known, fixed quantity. For instance, certain types of hazardous materials must be transported about the United States. Materials such as nuclear waste products, chlorine, or petroleum products must be transported to different locations for our industry to function. From society's or government's point of view, the risk issue of such transportation becomes one of judging which method of transportation is most acceptable. The transportation is necessary or we may say the gain is fixed. And there is always the possibility that some methods of transportation will be unacceptable given the invariant nature of the known gain.

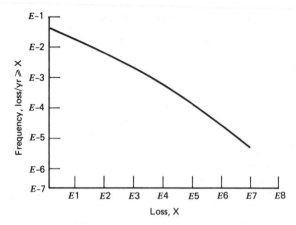

Figure 32.2   Risk spectrum example.

Judging the acceptability of risk may then become a problem of evaluating relative risk. Risk in a given system spreads across a spectrum of values from the frequently occurring minor accidents to the rare catastrophes. Some activities may generate a great amount of the total risk at lower levels of loss but at higher frequencies. Others may have their risk concentrated in a few major loss events.

A method of displaying risk in a manner that allows the judgment of relative risk is a risk spectrum. Figure 32.2 illustrates.

The figure represents a relationship between loss and frequency of that loss. It is read by entering the ordinate or frequency to find, for instance, that in this system one would expect a loss of 1000 or more about every 1000 years. The area under the curve of such a plot also represents the total risk of this activity in expected loss. This is the risk of only that loss dimension. If the loss dimension is deaths, then the area under the curve represents the total risk of death in a year of that system's operations. If the loss dimension is property damage, then a like risk is displayed. Total risk of a system can be displayed if the loss is the cost of all types of accidental loss.

A risk spectrum gives more information than merely the total expected loss. It also graphically displays the balance between the low and high loss levels in the operation of a system. Spectra with steep negative slopes have more of their risk at the lower loss levels or in accidents that result in lower losses. Flatter slopes display risks that have more of their total risk concentrated near the catastrophic levels. As there is a bias against catastrophic risks even through they may be rarely occurring events, flat risk spectra tend to be less acceptable even when total risk is less.

Another method of judging acceptable risk from its spectrum, is to compare the spectrum in question to one that is clearly acceptable. Natural risks are more likely to be acceptable to society than man-made. Thus, it may be helpful to display natural risk spectra on the same graph. Figure 32.3 is an example.

From Figure 32.3, a number of conclusions could be drawn, although note at the onset that we are dealing only with the risk of death. For instance, we might observe that of greater concern than nuclear power in a government's risk management, are earthquakes, tornadoes, and hurricanes. Another observation might be that since society accepts the risk of meteorites without protest it should accept the risk of nuclear power. That this is not the case is a function of risk perception. It is interesting to note from the figure that we can expect an accident causing 1000 deaths or greater every million years from either meteorites or the operation of 100 nuclear power plants. Figure 32.4 offers another set of risk spectra for consideration.

In this figure, we can see the rather startling greater risks of many human-caused events than from the operation of 100 nuclear power plants. More specifically, if society is concerned about nuclear power safety there should be a greater concern about chlorine releases. Figure 32.5 presents an approach to judging the risk acceptability of a narrow problem.

In this figure, we can see that the shipment of plutonium by whatever means poses significantly lower risks than chlorine shipments. We can see that the highest

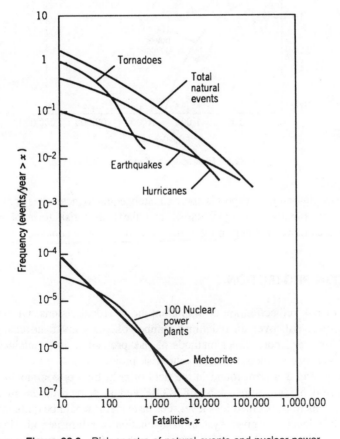

Figure 32.3 Risk spectra of natural events and nuclear power.

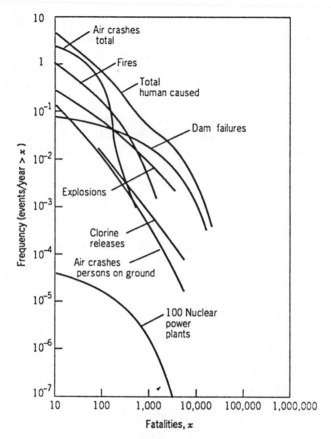

Figure 32.4   Risk spectra of man-caused events.

risk mode of plutonium transport is approximately equal to meteorites or 100 nuclear power plants. Finally, the figure shows that the lowest risk form of plutonium shipment is as oxide powder and that this risk should be acceptable.

32.5   RISK PROJECTION

Methods of risk projection are an important aid to risk decisions. Of course, risk spectra project risk over an indefinite number of years given no change in the system. However, more direct methods of risk projection are available. We will illustrate with one method, extreme value risk projection.

Return period is a term found in the field of risk, but not commonly in safety. Return period has the dimensions of MTBF and is defined as the mean period between events of a specified loss level. For instance, it would be quite informative to plot a loss level in a given system as a function of return period. Having done so, it is possible to predict a time during which a given level will occur even if such a loss has yet to occur in this system. If a system had yet to experience a loss level of $1 million but had a number of other accident losses with a maximum of $500,000, it would be possible to predict the mean number of exposure hours, months, or years between a loss level of $1,000,000 assuming that such a loss was possible. A return period of three years on such a loss level in a system two years old would indicate such a loss in the coming year.

Extreme values in specified periods of time exhibit an extreme value distribution behavior. The extreme value method may not be performed *a priori* and thus is not strictly a system safety method. It depends upon system data history to make the projections. However, it is quite useful when system safety is being applied to a deployed, operational system.

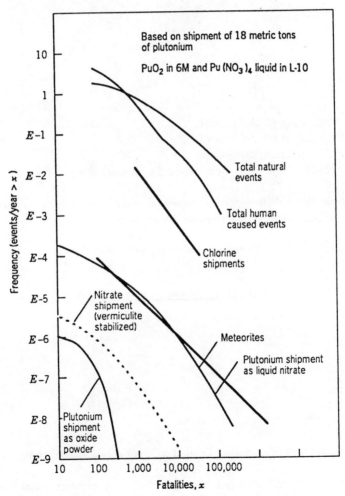

Figure 32.5 Risk of plutonium shipment by rail.

The extreme value is left skewed when the smallest values in a given period are used as the database, and right skewed when the largest values are used. In risk projections it would be expected that large values would be used.

The cumulative distribution function for the extreme value distribution is

$$F(x) = \int e^{-e^{-y}} \tag{32.4}$$

The probability density function for the extreme value distribution is:

$$f(x) = \alpha e^{-y-e^{-y}} \tag{32.5}$$

where

$$-\infty < y < +\infty$$

$$y = \alpha(x - \mu_0) \tag{32.6}$$

and

y = the normalized unit variable
x = the value of the random variable
μ_0 = the mode of the distribution
$1/\alpha$ = measure of the dispersion, called the Gumbel slope

The cumulative distribution function and probability density function for the logarithmic extreme value distribution are given by:

$$\Pi(x) = \int \exp -(\mu_0/x)^k \tag{32.7}$$

$$\pi(x) = -k(\mu_0/x)^{k-1} \exp -(\mu_0/x)^k \tag{32.8}$$

where

$$0 < x < \infty$$

and

x = the value of the random variable
μ_0 = the mode
$1/k$ = the measure of dispersion, called the geometric Gumbel slope

The extreme value distribution may be plotted on appropriately scaled paper that is commercially available. The lower abscissa is the cumulative percentage of the extreme values. The upper abscissa is the return period. The function that relates the lower abscissa to the upper is $1/(1 - p)$ where p is the cumulative probability. Thus, a cumulative probability of .9 will translate to a return period 10 years. A cumulative probability of .95 will translate to a return period of 20 years. If we gather data on the largest losses in a system in each year of its life and we find a .9 probability of having a given loss level or less, then the return period for that loss level would be 10 years.

The ordinate of such a plot will be the extreme value loss level. It may be a uniform or logarithmic scale. If a plot of the extreme losses in consecutive periods approximates a straight line on extreme value paper with a uniform ordinate, the accident causes are usually independent. This would indicate a strong safety program with control of common accident causes. This is because accidents caused by multiple independent causes will usually increase linearly with respect to the cumulative probability scale. If the plot of extreme losses approximates a straight line on logarithmic paper, there is a strong possibility of a common cause of accidents. Such a situation would indicate a revision of the safety program to find and eliminate common accident causes. This is because costs of accidents tend to increase exponentially with common causes.

For a more complete discussion of the method see King (1981). An example will illustrate the method.

Select a system exposure period in which the maximum accident loss will be recorded. In this example a year is selected. The maximum cost of an accident in each of the five preceding years of system operation is the following.

Year	Cost, $
1	5,930
2	88,850
3	10,070
4	497,000
5	38,000

Rank the selected accidents in order of increasing cost. Such a ranking assumes that the system has remained constant from the first period to the last. Calculate the cumulative probability of each accident cost by $N_i/(N + 1)$.

Cost, $	N_i	$(N_i/(N + 1))$
5,930	1	.167
10,000	2	.333
38,000	3	.500
88,850	4	.667
497,000	5	.833

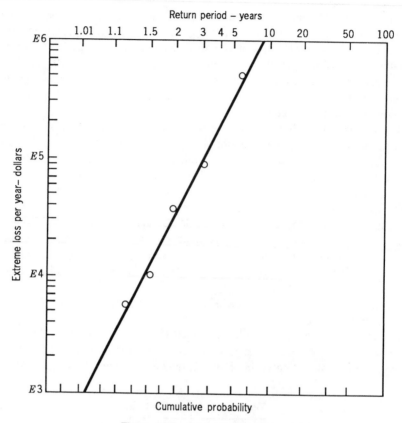

Figure 32.6   Extreme value plot.

We have determined the cumulative probability as a function of each period's maximum accident cost. The data may now be plotted on either logarithmic or uniform extreme value plotting paper. In this case, we will select the logarithmic paper. Figure 32.6 is the plot. Select the vertical scale that will permit extrapolation to a desired loss level. This should be no more than three times the maximum experienced loss.

Note that we can extrapolate easily to a return period of 10 years for a $1 million accident. The plot also facilitates finding return periods for lower loss accidents, either experienced or not.

32.6   RISK ASSESSMENT AND MANAGEMENT

The process of risk assessment and management in a developing system can be examined as a set of definitive activities.

1. *Identify and Evaluate Hazards.* There are a number of techniques identified earlier in this book that will enable the safety professional to identify hazards.

Having identified system hazards, they must be evaluated. There are two components to the hazard evaluation process—likelihood and consequences. Both must be completed for the majority of system hazards.

2. *Assess Risk.* Assessing system risk involves evaluating the expected loss or risk of each hazard and summing over system operations and operational life. The evaluation should consider the low-consequence, high-frequency hazardous events as well as the low-frequency, high-consequence events. The assessment may be multidimensional or in units of value. If multidimensional the risk may be stated in terms of expected deaths, injuries, workdays lost, or property or environmental damage.

3. *Develop Countermeasures.* Control of risks must commence with selection of sets of system design changes, procedural changes, operational constraints, or operator improvements that will reduce the likelihood or severity of the accident events. However, it is unlikely that countermeasures will be successful in reducing risks to zero. The countermeasures should control risk to an acceptable level.

4. *Allocate Resources.* Countermeasure implementation, even in the design phase of system development, will require additional resources above those previously budgeted. If resources are not available, the risk assessment activity becomes one of "pointing a finger" at the problem. If countermeasures exceed the always limited resources, then new sets of countermeasures must be developed.

5. *Accept Residual Risk.* Having decided on the resources and thus the countermeasures that will be implemented to control system risk, the remaining risk of system operation must be overtly accepted. This acceptance must be under the purview of management, not the safety professional. The risk assessment report must be presented to management in a form that highlights residual risks. The responsibility for acceptance of these risks must be clearly established. It is the responsibility of the system safety professional to appraise management of the risks remaining in the system following implementation of countermeasures. It is these risks that must be accepted, or if unacceptable, be further mitigated.

32.7 SAFETY VALUE ANALYSIS

It is the rare safety practitioner who does not feel that insufficient resources are being expended on safety in his or her responsibility domain. However, such feelings should be accompanied by an understanding that additional resources expended on safety should return more than their value if the uncertainties of such an expenditure are to be worth accepting. The problem must be faced jointly with advocates from other areas of system performance as they also attempt to justify additional resources for improvements in their system responsibility areas.

Methods that adjust value expended for value received must resolve the multidimensional nature of hazard consequences into a single dimension of value. This may be a form of utility value or simply dollars. Ratios of effectiveness may be formed to measure the relative worth of the various calls on the limited resources available to improve the system. A logical process is needed that equates value gained in safety improvement on the margin, to value expended.

32.7.1 Value Analysis Process

The value analysis process in safety simply refers to an interactive decision methodology that relates the elements of the decision process. The process focuses on differences between safety goals as stated in system specifications and the level of safety attainable within the twin constraints of technology and resource availability. Figure 32.7 describes the process in conceptual form.

The process commences with a statement of safety goals in appropriate dimensions. A goal might be stated as no fatalities during the system life. Although such a goal would appear noble and appropriate to the lay person, safety analysts are aware that fatalities are a stochastic variable. Thus, it will be difficult to design a system to make fatalities impossible. Therefore, the no-fatality goal must be accompanied by a probability or confidence. It would be more appropriate to state the fatality goal as a function of a fatality rate. This would tie the fatalities to exposure and allow the introduction of likelihood into the fatality goal. As more systems are placed in service, a fatality rate goal could continue to be a proper measure. The goal of no fatalities in the operation of a space station would become questionable as more stations are placed in orbit with, perhaps, hundreds of people exposed to their hazards.

In a similar manner human injury goals can be established. Here, it must be recognized that both the number of injuries and the severity of those injuries is a variable. Therefore appropriately stated goals for injuries, must be expressed in terms of numbers and severity of injuries as well as the likelihood. Again a rate is the most convenient way of grappling with such variables.

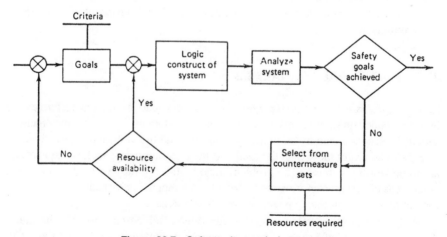

Figure 32.7 Safety value analysis process.

Safety goals may also be stated in terms of costs. One may try to design into a system an upper bound on costs of accidents. Again such costs are variable. They will also be strongly dependent upon exposure. A dimension in which a cost goal may be stated that would consider the probabilistic nature of such costs is the *expected cost*. Expected cost must be related to some exposure period or calendar period. It would be proper to use expected cost per year or even expected cost per operating or exposure hour. An expected cost goal can contain elements of death, injury, and property damage. It is coupled with likelihood. Thus, expected cost as a system safety goal is the most encompassing and most rational. While any safety goal might be termed a risk goal, a safety goal stated as expected costs would fulfill the necessary requirements of risk description.

Having established a goal, a system safety analytical model must be developed. This model must be only as explicit as necessary to verify the goals. In the case of a large, high-technology system, the model would need to be in the form of fault trees, block diagrams or reliability models of the many subsystems. These would then need to be drawn together into a master system model that could validate the stated safety goals. In a small system, perhaps a safety helmet, the model could be quite simple.

The completion of a system safety model will allow the analyst to gather data on system components and exercise the model. The output of the modeling process will be compared to the system safety goals. If the goals are met, the system is in a satisfactory state of safety and development may proceed. If not, modifications must be made. The modification process is one of examining system redesign options, and, having selected some design modifications, reexercising the system model. This process may be thought of conceptually as selecting sets of safety countermeasures. There may be several sets of such countermeasures that offer hope of achieving the stated safety goals. However, all countermeasure sets have a cost attached. So the modification process becomes one of first examining the technology for countermeasures, and then determining if resources are available to implement these countermeasures.

Figure 32.7 implies that affordable countermeasures will be exercised in the system model as a search is made for a state of safety that will meet the goals. If this process fails, than the goals must be modified or the system is unsatisfactory.

The safety value analysis as depicted in this figure may be done formally or informally. However, such a process must be addressed, more formally on large, complex systems, or a mismatch between the desired state of safety and that achieved will occur.

32.7.2 Expected Value

The calculation of expected value of the large number of stochastic loss variables present in a complex system, can be quite complex. We have previously examined a concept of expected value as a set of conditional probabilities (equation (32.3)). A more concise and usable concept of expected value is equation (32.9).

$$\overset{i}{\Sigma} E(V_i) \ = \ \int P_i(x) V_i(x) dx \tag{32.9}$$

where

$P_i(x)$ = Probability density function of the ith accident
$V_i(x)$ = Utility value function of the ith accident (negative)
n = Total number of accidents

In most cases, it would be impractical to develop and integrate equation (32.9). A suitable approximation that lends itself to calculation is equation (32.10).

$$E(V_i) \ = \ \overset{i}{\Sigma} P_i V_i \tag{32.10}$$

This equation may be evaluated from the Poisson distribution of Chapter 13. The Poisson infinite series for the probabilities of 1 through n accidents, where $n = \infty$, is the following:

$$\frac{e^{-\lambda t}}{0!} \ + \ \frac{\lambda t e^{-\lambda t}}{1!} \ + \ \frac{(\lambda t)^2 e^{-\lambda t}}{2!} \ + \ \frac{(\lambda t)^3 e^{-\lambda t}}{3!} \ + \ \cdots \tag{32.11}$$

Multiplying this equation by the utility values of the numbers of accidents 1 through n, and assuming that 2 accidents of the same consequences have twice the value of 1:

$$\lambda t e^{-\lambda t}(V(1)) \ + \ \frac{(\lambda t)^2 e^{-\lambda t}}{2!} \ (2V(1)) \ + \ \frac{(\lambda t)^3 e^{-\lambda t}}{3!} \ (3V(1)) \ + \ \cdots$$

Factoring out common factors:

$$\lambda t e^{-\lambda t} V(1) \left(1 \ + \ \lambda t \ + \ \frac{(\lambda t)^2}{2!} \ + \ \cdots \right)$$

$$= \ \lambda t e^{-\lambda t} V(1) e^{\lambda t} \tag{32.12}$$

$$= \ \lambda t V(1)$$

Equation (32.12) is the expected value of all accidents that may occur of similar consequences. Summing over all possible accident consequences will provide the total expected cost for system exposure, t.

32.7.3   Cost of Safety

A safety manager concerned with selection of a state of safety at which his or her system should operate will search for a criterion that will measure safety performance in such a way that the optimum state will be indicated. It would be helpful if the safety manager could present his criterion and its associated logic in a way that project management could understand and accept. Figure 32.8 depicts two basic types of safety costs and may suggest an optimum point for the state of safety.

Examining this figure, we can see that the two types of safety costs are presented along with a curve of the total cost. Consider first the cost of accidents scale at a state of zero safety. If a system is operated at this state the cost of accidents will be high, but finite. All elements of the system will be lost. There will be personnel deaths and serious injuries. The indirect costs of these accidents will also be high but calculable. It is, of course, not a desired state of safety, although no costs are being expended on accident prevention or countermeasures.

It is clear that at a perfect state of safety there will be no accidents and thus no costs associated with them. It may be that the relationship between costs of accidents and degree of safety is linear. This would mean that at a 50 percent degree of safety the costs of accidents would be 50 percent of the costs at zero safety.

Turning to the cost of countermeasures scale at a perfect state of safety, we can see that the cost is infinite. Some might argue that such a state for a complex system cannot be achieved. As this function moves toward a complete lack of safety, we can see that it is nonlinear but approaches the origin of the plot. The desired state of safety must lie somewhere between these two extremes.

By adding these two cost curves the total cost of safety is developed. As might be expected, this curve has a minimum near the point at which the accident and countermeasure costs are equal. It need not be precisely at this point. The location of the minimum point will be determined by the nature of the two supporting curves near the crossing region.

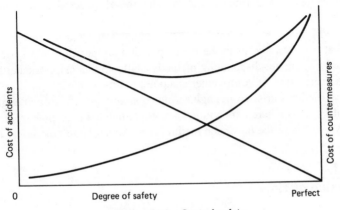

Figure 32.8   Cost of safety.

The two categories of cost may be divided into direct and indirect. The direct costs of accidents will include mortality, injury, morbidity, and property damage. Indirect accident costs would be primarily loss of individual productivity, loss of system productivity, insurance and costs of litigation. It is quite possible that the indirect costs of accidents will exceed the direct costs.

Direct countermeasure costs will include design changes principally for safety, safety personnel, safety systems, safety education, and training programs. The indirect countermeasure costs may be restrictions on system operation to enhance safety.

Examining the total cost curve for the optimum state of safety, the minimum is an obvious choice. It is difficult to imagine a reason to operate at a state less than minimum. However, many of our early aerospace systems attempted to do so at an exorbitant cost. If one wishes to operate at a safety state entailing a total cost greater than the minimum, it is possible to justify. It the costs of accidents curve has a steeper negative slope than we have imagined, the true minimum cost point will be forced to the right on the plot.

Large, complex, politically sensitive systems in which a single accident may become a catastrophe, are systems in which the slope of the accident cost curve may always be steeper. Examining nuclear power, we see such a system. A single accident has done greater damage to the entire industry than could possibly have been imagined prior to its occurrence. In this industry, it is clear that the minimum cost point is well to the right of where the industry was operating. A prototype aircraft system is another example. In such a case, the crash of the prototype may result in the cancellation of the program.

Such steep negative slopes of the accident cost curve justify the expenditure of greater resources to prevent the catastrophic accident. Once the accident emerges from Pandora's box there is no putting it back. The owners and operators of the system and the entire related industry must contend with an outraged public and a lack of funds to support further development. This unpleasant and costly situation may threaten the fiscal stability of the parent organization.

We must now note that the cost of safety curves are at different points in time. We expend accident countermeasure costs today to reduce accident costs tomorrow and over the life of the system. Thus, we see that these curves are not a precise model that will guide us to a level of safety expenditure, but a conceptual model of losses that we must accept in the development and operation of a system.

Fortunately, there are discounting methods available that can evaluate the value of time and bring these curves together at a point in time in the system development cycle. These methods arise from engineering economy and are used in engineering or technical systems where expenditures are contemplated over time, that must be justified. We will describe how such methods may be applied to the safety decision problem.

32.7.4   Value of Time

In safety analysis, we are continually faced with the problem of assessing the uncertain future. Merely to predict event occurrences in the future is difficult enough, but in safety we must compare the values of these events with expenditures that may be made today to prevent *some* of those future events occurring with some frequency or severity. The decision to accept a current state of safety in a contemplated system or to recommend further expenditures to alter this state, must be based on engineering and economic analysis of the system today and then tomorrow. The judgments are both technical and fiscal.

The concern in safety is simply acceptability, both for the user and public at large. Both must accept the system and its attendant losses for the system to be a success. As was seen in Figure 32.8, the cost of safety increases exponentially with a need for improvement. In the final analysis we are seeking a balance between the capacity of the system to perform its mission satisfactorily and acceptability of the losses that result from completion of this mission.

In safety exactly as in other engineering decisions, fundamental questions must be answered.

1. Why do this at all?

Are the proposed design changes necessary?

Will they be effective in reaching a needed level of safety?

Do we have the resources to fund such a project?

Will the customer support such an effort?

2. Why do it now?

If the safety change is implemented later in the life of the system, will it be equally effective?

If implemented later will this incur significant costs increases?

Have we passed the minimum cost point for implementation of safety countermeasures?

3. Why do it this way?

Have we examined all of the countermeasure sets that will provide an equal or greater degree of safety improvement?

Have we examined the cost implications of our current plan and if so are the costs acceptable?

Are the countermeasures compatible with system safety goals?

Has the value of time been considered in the development of costs and benefits of the set of countermeasures we are about to implement?

These are only a few of the pertinent questions that must be addressed in evaluating safety countermeasures. The value of time is a well known consideration in investment decisions of all kinds. It is, perhaps, of even greater importance in safety investments. Time has value whether or not the potential investor chooses to recognize its impact. Clearly, any organization that has surplus funds to invest has alternatives that will return more than the value invested at some future point in time. All investments with future payoffs have risks. An increment of the future return must compensate for risk.

Beyond the risk compensation requirement, each investment made is, in effect, foregoing the opportunity to invest in a safe, relatively risk-free investment that will return a reasonable rate of return. Each individual and organization in the United States has the opportunity to make such a risk-free, liquid investment. For individuals, this investment may be a passbook savings account. For a large business, this *opportunity foregone* to invest in safety may be Treasury bonds.

Each investment being considered by companies or individuals must be compared to the opportunity foregone or, as it is sometimes called, the *opportunity cost*. If an investment is to be considered worthy, it must provide a rate of return greater than the opportunity cost. An investment in a project because of a subjective feeling of need is always a mistake. Nevertheless, this approach is common in safety investments. Our concern as safety professionals should be to command resources for safety improvements only when they can meet a strict criterion of productivity. This means that the rate of return on the safety investment must exceed the opportunity cost and rank well with other possible investments that can be made in the system. If such is the case we will be in a position to defend it against all challenges as well as justify it to ourselves and colleagues. And if we use sound engineering economic principles to judge our safety investment we will be presenting a defense understood by all.

Our approach in safety investment analysis will be to show that future values saved in accidents prevented or mitigated, will, when discounted to the present, have a greater value than the costs of the safety countermeasures when similarly discounted. And the discount rate will be the opportunity cost or, if the investment capital is borrowed, the cost of the borrowed capital.

32.7.5   Engineering Economic Factors

Engineering economic factors are methods of discounting to the present or compounding to the future, investment values. The factors are simple multipliers of value amounts that the analyst may wish to displace in time at some compounding or discounting rate. There are a relatively small number of these factors with which every type of compounding or discounting problem can be solved:

Single-payment and series present worth factors

Single-payment and series compound amount factors

Capital recovery factors (considered as a series)

Sinking fund factors (consider as a series)

A single-payment factor means that a single value is discounted or compounded. A series factor means that a series of equal annual values are discounted or compounded.

The theroetical development of these factors may be found in any good work on engineering economy. In this brief presentation only their use will be explained.

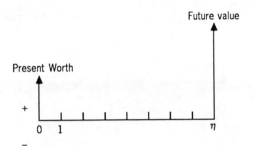

Figure 32.9 Single-payment present worth factor.

Present worth factors (sometimes called present value) are factors that, when multiplied by a future value or series of values, will compute the present equivalent value, considering the value of time, to the planning horizon as measured by the discount rate. The single payment present worth factor will find the present worth of a single value received at some future point in time. This factor will discount this future value at some compound discount rate, either the opportunity cost or the cost of borrowed capital. Figure 32.9 illustrates.

$$\text{Single payment present worth factor:} \quad \frac{1}{(1 + i)^n} \quad (32.13)$$

where

i = annual discount or interest rate
n = number of periods, years if i is annual rate

The series present worth factor will, when multiplied by the value of one of a series of equal values received at equal periods over the future to the planning horizon, find the present worth of those values. Figure 32.10 illustrates.

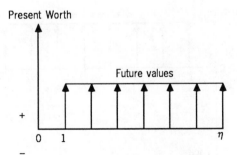

Figure 32.10 Series present worth factor.

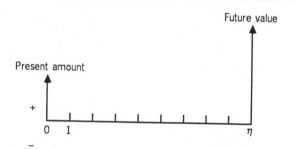

Figure 32.11   Single-payment compound amount factor.

$$\text{Series present worth factor:} \quad \frac{(1 + i)^n - 1}{i(1 + i)^n} \qquad (32.14)$$

Compound amount factors are factors, that when multiplied by an investment made today, or a series of investments, will give the future value of the investment to some specified planning horizon over a given number of compounding periods at a given compounding rate. The single-payment compound amount factor will compound a single investment made today over some future number of compounding periods to find the future value. Figure 32.11 illustrates.

$$\text{Single-payment compound amount factor:} \quad (1 + i)^n \qquad (32.15)$$

The series compound amount factor will, when multiplied by a series of equal investments made over equal periods to some future planning horizon, find the future value of that series of values compounded at a given interest rate.

$$\text{Series compound amount factor:} \quad \frac{(1 + i)^n - 1}{i} \qquad (32.16)$$

Capital recovery factors will, when multiplied by an amount invested or loaned, determine the series of equal payments necessary to repay the principal amount and with compound interest over a specified number of periods. Figure 32.12 illustrates.

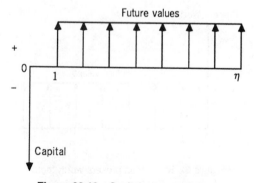

Figure 32.12   Capital recovery factor.

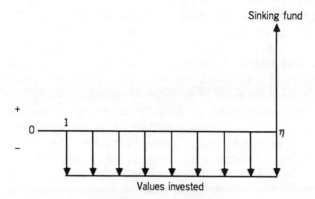

Figure 32.13   Sinking fund factor.

$$\text{Capital recovery factor:} \qquad \frac{i(1 + i)^n}{(1 + i)^n - 1} \qquad (32.17)$$

Sinking fund factors will, when multiplied by a value desired to be available at some future planning horizon, give the series of equal annual payments that must be invested at a given interest rate for the specified number of periods, to provide the desired future value. Figure 32.13 illustrates.

$$\text{Sinking fund factor:} \qquad \frac{i}{(1 + i)^n - 1} \qquad (32.18)$$

32.7.6   Value of Time Calculations

If we consider safety investments in the light of the value of time, we can establish a method by which the worth of the investment can be evaluated. Figure 32.14 illustrates the logical relationships of the values in question.

Figure 32.14 shows a value, b, as the annual cost of accidents prior to installation of safety countermeasures. We will assume that a countermeasure will reduce the

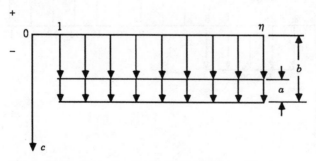

Figure 32.14   Evaluating the worth of a safety countermeasure.

annual cost of accidents in a given system by the amount, a. The cost of this countermeasure is shown as, c, expended at present. The worth of this safety investment, or more precisely the net present worth, will be the net difference between the present worth of the annual accident costs saved by the countermeasure and the present cost of the countermeasure. The series present worth factor will be used to find the present worth of the series of savings.

In this simple example, we see the difference between safety investments and the more traditional form of investment analysis. In the safety case we invest in countermeasures to reduce accident costs over the life of the system. This differs from the traditional investment that produces future cash flows. Investments in safety reduce future costs rather than generating positive cash flows. We will examine other examples to illustrate the calculations.

Assume an engineering change to a system for improved safety with a present cost of $220,000 and an annual cost over five years of system development of $50,000. Assume further that we can calculate that the change will reduce annual system accident costs from $360,000 to $245,000 over a ten year operational life following the development cycle. Assume the opportunity cost of this investment is 10 percent. Figure 32.15 illustrates this problem.

Prior to deciding if this modest annual savings in accident costs of $115,000 is worth doing we must calculate the net present worth (NPW) of this safety investment. Note that in the above diagram of this calculation the $115,000 of annual savings is shown as positive. The calculations are made as follows.

$$\text{PW of } \$115,000 = \$115,000 \text{ (PW} - .10 - 10)(\text{PW}' - .10 - 5)$$

where
$(\text{PW} - .10 - 10)$ = Series present worth factor, 10 percent, 10 years

$(\text{PW}' - .10 - 5)$ = Single payment present worth factor, 10 percent, 5 years

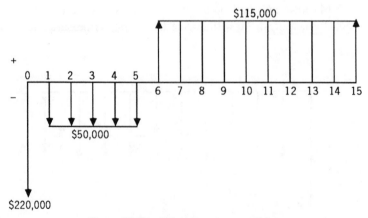

Figure 32.15 Safety investment problem.

$$(PW - .10 - 10) = \frac{(1 + .10)^{10} - 1}{.1(1.1)^{10}} = 6.1446$$

$$(PW' - .10 - 5) = \frac{1}{(1 + .10)^5} = .6209$$

PW of $115,000 $= \$115,000(6.1446)(.6209) = \$438,750$

PW of $220,000 $= \$220,000$

PW of $50,000 $= \$50,000(PW - .10 - 5) = \$50,000(3.7908)$

$= \$189,540$

NPW of investment $= \$438,750 - \$220,000 - 189,540 = \$29,210$

Thus, it appears that at a 10 percent opportunity cost this investment in improved safety is worth considering. Still to be answered is a question concerning the uncertainty of this return of $29,210. If conservative values have been taken throughout the calculation, the analyst may feel comfortable with the uncertainty. Before resources will be given to such an investment the return must be ranked with the returns of other investments in this system or within the organization that wish a share of the always limited resources.

If, perhaps, the capital must be borrowed to make this safety investment and the cost of borrowed capital is 13 percent, the calculation may have a different outcome.

PW of $115,000 $= \$115,000(5.4262)(.5428) = \$338,710$

PW of $220,000 $= \$220,000$

PW of $50,000 $= 50,000(3.5172) = \$175,860$

NPW of Investment $= 338,710 - \$220,000 - \$175,860 = -\$57,150$

Such an outcome will halt further consideration of this investment. We find that as the discount cost increases, the attractiveness of investments decreases, given the same fundamental values.

There may be a question as to how the $115,000 per year saved by the countermeasure, is found. The annual expected accident loss of a particular type is:

$$\lambda_i t(V(1)) - \text{expected annual loss of } i\text{th level} \qquad (32.19)$$

where $\lambda_i =$ the annual accident rate of ith level.

The countermeasure under consideration will reduce either the rate, λ_i, the value, $V(1)$, or both. Careful analysis can determine this annual amount.

Such methods of evaluation of safety investments do not consider what may called the *irreducibles* of the investment. These would be considerations of pain and suffering that accidents cause and upon which a value cannot be placed. In the category of irreducibles we must place the reputation of the organization. Harm to the reputation of an organization caused by a prominent accident, can have great negative value. It may be possible to evaluate irreducibles by utility value methods.

EXERCISES FOR PART V

1. Imagine a case where death will be accepted as an element of risk. Draw a hypothetical risk spectrum for this case. Examine your reasons for this position.

2. Compare the risk spectrum of an industrial organization to that of a military aviation organization.

3. Draw hypothetical risk spectra for the following systems. Compare spectra for acceptability.
   (a) Risks in the home
   (b) Traffic risks
   (c). Hang-gliding risks

4. Why is it necessary to discount future costs saved to evaluate the state of safety of a system?

5. Define opportunity cost.

6. Why should the analyst discount future safety costs saved at the cost of capital if the capital is borrowed to implement safety countermeasures.

7. What is the NPW of a safety investment of $5000 to reduced accident costs $700 each year for the next 10 years? The cost of borrowing the $5000 is 10 percent.

8. If the cost of borrowed capital in problem 7 is 6 percent, what is the NPW? Is this a wise investment in safety?

9. A new system is being developed. The development process will take 10 years. After deployment, the system is planned to be used for 15 years. A safety countermeasure is proposed that will save $4.25E6 per year in accident costs for the system operational life. This countermeasure will cost $2.5E6 at the start of development, $1E6 per year during development, and $2E6 at the time of system deployment. The opportunity cost of this investment is 13 percent. What is the NPW of this investment in safety? Is this a wise investment?

10. There is a possibility of investing in the safety of a new system to reduce the system failure rate of $2.7E-5$ per hour by 30 percent over the system life of 10 years. The loss per failure of this system will be $1E6. The cost of this safety countermeasure is $5E5. The system will be used $E4$ hours per year. The opportunity cost of this investment is 12 percent. What is the NPW of this investment in safety? Is it a wise investment?

11. If in problem 10 the cost of money use could be reduced to 9 percent, would this change the attractiveness of the safety investment?

REFERENCES AND BIBLIOGRAPHY

Block, M. K. and Lind, R. C., "Wealth Equivalent, Risk Aversion and Marginal Benefit from Increased Safety." Report prepared for Office of Naval Research, Arlington, VA by U.S. Naval Post Graduate School, Monterey, CA, AD 781375, May 1974.

Diamond, P. "Economic Factors in Benefit-Risk Decision Making, Perspectives on Benefit-Risk Decision Making." Report on a Colloquium on Benefit-Risk Relationships for Decision Making conducted by the Committee on Public Engineering Policy, National Academy of Engineering, Washington, DC, April 26–27, 1971, 1972.

Grant, E. L., *Principles of Engineering Economy*. The Ronald Press, New York, 1964.

King, J. R., *Probability Charts for Decision Making*. TEAM, Tamworth, NH, 1981.

Mao, J. C. T., *Quantitative Analysis of Financial Decisions*. The Macmillian Co., London, 1969.

Nertney, R. J., Clark, J. L., and Eicher, R. W., *Occupancy-Use Readiness Manual—Safety Considerations*. ERDA-76-45-1, *SSDC-1*, System Safety Development Center, 1975.

Newman, D. G., *Engineering Economic Analysis*. Engineering Press, San Jose, CA, 1980.

Otway, H. J., Pahner, P. D., and Linnerooth, J., *Social Values in Risk Acceptance*. International Institute for Applied Systems Analysis, Laxenburg, Austria, 1975.

"Reactor Safety Study," An Assessment of Accident Risks in U.S. Commercial Nuclear Power Plants," WASH-1400, 1975.

Roland, H.E., "Risk and Safety Value Analysis, *Professional Safety*, 26–29, September 1983.

Roland, H. E., "The System Safety Domain," *Hazard Prevention*, March/April 1988.

Rowe, W. D., *An Anatomy of Risk*. Wiley, Sons, New York, 1977.

Wilson, R., The Costs of Safety," *New Scientist*, 68(973), October 20, 274–275, 1975.

VI

DECISION ANALYSIS

33

DECISION METHODS
FOR SAFETY

A life of satisfaction and reward in our complex society is guided and directed by a continuous thread of decisions. Each of these decisions will return its benefit or penalty at some point in the uncertain future. Almost every action we take, with the exception of involuntary responses, results from a decision and each decision is accompanied by some risk. The risk is a possible undesired outcome of the decision as the future unfolds. The decision as to whether to drive a given route to work may result in an accident or an avoidance of one. At the minimum there will be a payoff of reduced driving time or the cost of a greater time wasted in an automobile. We now know that a decision to smoke, formerly a nearly involuntary action, now carries risk. The major decisions of our life such as our mates, our homes, our jobs, and our life-styles are fraught with risk but offer hope of benefit. Safety-related decisions in the development and operation of complex systems clearly offer potentials for payoffs and the threat of losses.

Herein we will present a few of the fundamentals of decision-making under *uncertainty*. Uncertainty in this sense means that the future environments in which the system must function carrying with it decisions of the past, are unknown. They are, however, definable as a limited set of alternative futures.

Decision making under certain is also necessary, but much simpler. For instance, if one *knows* it is going to rain one must decide to carry an umbrella. The decision may seem an obvious one, but it must be made and carried out or one will incur the loss of being out in the rain without protection.

If one is making decisions under uncertainty concerning the development of complex, high technology systems, there are many alternative futures that are of concern. Technological innovation and discovery may be a nearly all-important future state that must be considered. The desires of the customer are important and, at times, difficult to predict. Funding levels are most important. Policies that will influence the development and employment of the new system are important considerations. Competitive products, their development and state, must be considered. The uncertain futures referred to herein will be sets of levels of these areas of influence and concern. Before a rational decision can be made concerning a given system, sets of these influential futures must be formulated as structure and background for the decision process. Let us take a simple example.

Assume we are concerned with an oil-drilling investment. Of course, we do not know if we will strike oil. If we knew, it would be a problem in decision making under certainty and a much simpler decision. In this case, assume we own the oil lease. We can calculate that if we invest in oil drilling with an investment cost of $3E5$ and strike oil, the net payoff will be the return from the oil of $2.3E6$ minus the drilling cost of $3E5$, or $2E6$. If we do not strike oil, the investment will result in a straight loss of $3E5$, neglecting tax implications. If, on the other hand, we do not invest in drilling we may sell our oil lease for $E5$ with a contingency fee of an additional $2E5$ if there is oil present. In this rather simple case with only two futures we can quickly see that the no investment decision results in positive payoff in both futures, oil or no oil. If we do invest, we may lose $3E5$. If this loss is unacceptable we have no choice but to decide not to invest and forego the possibility of a $2E6$ payoff. Even if we can accept the possible loss in the investment decision, there may remain a question as to which decision offers the greatest payoff considering both possible future states. To evaluate this possibility we must determine the probability of oil and thus no oil, and place the outcomes in a decision matrix for analysis. We will assume that geologists tell us the probability of oil is .2. Table 33.1 is such a matrix.

The expected payoff from the decision Invest is $.2(\$2E6) - .8(\$3E5) = \$1.6E5$. The expected payoff from the decision No Invest is $.2(\$3E5) + .8(\$E5) = \$1.4E5$. The Invest decision appears to be slightly preferred assuming that the large probability of having a loss of $3E5$ is acceptable.

The elements of the decision-making process under uncertainty are thus the following.

1. What are the performance parameters of interest in the operation of our system?

Table 33.1. Decision Matrix for Oil Investment Future

Decision	Oil $P(\text{Oil}) = .2$	No Oil $P(\text{Oil}) = .8$
Invest	$2E6$	$-\$3E5$
No invest	$3E5$	$E5$

2. How will these performance parameters vary under the possible states of the uncertain future?

3. What value do we place on the magnitudes of the performance parameters?

4. How will the values we place on the performance parameters vary under the states of the uncertain future?

5. In view of the answers to the above four questions, as they apply to the alternatives under consideration, which of these alternatives shall we choose?

The scope of these five questions is beyond the example problem above. However, we will eventually address them all.

33.1 DELPHI AND RELATED METHODS

The plural judgment of the committee approach to decision making has its roots in the Delphi technique. The Delphi method, once widely acclaimed as the answer to decision making under uncertainty, has now been largely discredited by its originator, the Rand Corporation. However, because of its historical significance, it is worth reviewing.

The site of Delphi in ancient Greece was believed to be sacred because of hot springs in the area. It was dedicated to Apollo. A group of wise men built a temple on the site and developed a reputation as a place where the Gods could provide the answers to important and difficult questions concerning the uncertain future. The wise men, or priests, trained a local woman to be the oracle known as Pythia. The well-to-do traveler arrived with a question for decision and was detained for a few days, revealing, in the interim, his question to the priests. In the mean time he refreshed himself with good wine and food in pleasant surroundings. After an appropriate period, the traveler was ushered into the presence of the oracle where the question was again stated. The oracle uttered gibberish that was duly interpreted by the priests as the answer to the traveler's question. As the priests were wise men and knowledgeable in many fields, the passage of time found them correct more often than not.

The site and the priests gained a reputation for wise decisions and each of the major city-states of Greece built a temple at the location. Governments sought out the priests for solutions to disputes among the warring cities. The city and its priests served as an early supreme court.

The original Delphi decision-making method as advocated by Rand was a committee method with a unique twist. The committee was separated so that the members would not be influenced by an informal leader. The committee would be polled by the administrator for their decision. Their independent responses were statistically assembled. A limited set of information concerning the responses of the group was fed back to the committee for a second iteration of the process. It was once believed that the mean of the committee would migrate to an approximation of the true value with three iterations.

33.2   COMMITTEE DECISION METHODS

There are a variety of committee decision techniques. The *consensual* method involves a relatively small, perhaps no more than a 10-person committee. The committee meets together and the question for decision is presented by the leader or facilitator. The method allows for unlimited discussion of the question, its ramifications and ultimately focusing on the decision. The facilitator or leader guides the group attempting to avoid paralysis on a minor point or personal animosities. Informal leaders arise and are allowed to influence the decision. The goal of the method is agreement on the decision by each member of the committee. There are no parliamentary rules of order and no voting. When successful, the method has the strength of unanimous agreement. The disadvantages are time consuming and no certainty that a decision can be reached.

The *parliamentary* is the most common committee approach. In this method, the committee meets together under an appointed or elected leader. The committee may be larger than in the consensual approach. Robert's parliamentary rules of order prevail and presentation of motions and discussion of questions are formally governed by them. Eventually, a vote is taken with the majority making the decision. The method has the advantage of almost assuring a decision can be reached. The decision will be subject to political maneuvering of power groups within the committee. Another disadvantage is that the losing voters may not support implementation of the decision.

The *survey* decision is sometimes called Delphi. This approach frequently follows a research study of a question or issue. In this method a questionnaire is sent to a group of experts on the subject in question. If research has been completed the results of the research are also sent for consideration. The experts are asked a set of questions concerning the question for decision, and, if research is available, questions concerning the results of the research. The answers are tabulated and used to support or deny the results of research.

33.3   METRIC MATRIX HIERARCHICAL ANALYSIS

Determining the weights to be applied to safety design factors when data is sketchy or subjective, continues to be a problem in safety analysis. This problem arises where the system being designed is new technology or rate information on components is sketchy. Metric Matrix Hierarchical Analysis (MMHA) is a method of coping with many subjective factors in safety analysis.

MMHA is a mathematical technique utilizing the computation of dominant eigenvalue and eigenvectors of a metric matrix. The method facilitates the transformation of subjective judgments of experts to quantified values. In addition it provides a consistency check on those subjective judgments. Using this method subjective judgments of experts may be used to determine the relative importance of design criteria. MMHA can determine those areas of design activity requiring attention to improve the safety of a new design. While the method requires matrix manipulation, computer programs will provide ease of use.

The method is useful for:

1. Predicting outcomes.
2. Selecting alternatives.
3. Allocating resources according to priorities.
4. Conducting cost/benefit comparisons.
5. Planning projected or desired futures.
6. Risk management.

The method consists of (1) breaking down a complex unstructured situation into its component parts, (2) arranging these parts or variables into a hierarchical order, (3) assigning numerical values to subjective judgments on the relative importance of each variable, and synthesizing the judgments to determine which variables have the highest priority and should be acted upon to influence the outcome of the situation.

The steps in the MMHA are:

1. Define the problem in detail.
2. Structure it in a hierarchy of levels of detail. Start with the highest hierarchy which defines the overall objective. Continue with the levels of hierarchy until the lowest hierarchy is determined which includes the final actions and alternative plans that would contribute to the main objective. The hierarchy depicted graphically will then show the interdependence of variables in the problem.
3. The relative importance of all variables is determined by subjective (expert) judgment and assigned numbers from 1 to 9. If there is a committee providing evaluation, the variable with the highest numerical value (highest priority) becomes the best choice.
4. Since subjective judgment is not a perfect procedure, a test must be made to ensure logical consistency. This is done by computing the Consistency Index (CI) and the Consistency Ratio (CR).

Figure 33.1 and Table 33.2 illustrate hierarchical levels and importance values of the variables.

Having identified the alternative outcomes and assignment of variable importance factors by the experts, pairwise priorities among the variables are established that will influence relations between variables such that:

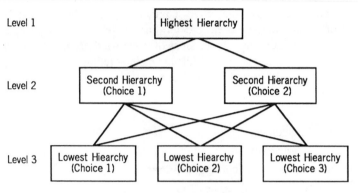

Figure 33.1 Hiearchical structure.

$$a(i,j) = W(i)/W(j) \qquad \text{for } i,j = 1, 2, \ldots n \qquad (33.1)$$

A matrix A is formed of these weighting factors $a(i,j)$.

$$A = \begin{bmatrix} W(1)/W(1) & W(1)/W(2). \ . \ . \ . \ .W(1)/W(n) \\ W(2)/W(1) & W(2)/W(2). \ . \ . \ . \ .W(2)/W(n) \\ \vdots & \vdots \qquad\qquad\quad \vdots \\ W(n)/W(1) & W(n)/W(2). \ . \ . \ . \ .W(n)/W(n) \end{bmatrix}$$

The matrix A is multiplied by weights, W, resulting in a matrix nW. From the matrix a system of homogenous linear equations must be solved, that is, $Aw = nw$ or $|A - nI|w = 0$. A nontrivial solution exists if and only if the determinant of $|A - nI| = 0$. The condition that the determinant equals zero leads to an nth degree equation in λ called the characteristic equation. The roots $\lambda(i)$, $i = 1, \ldots n$ of the characteristic equation $|a - \lambda I| = 0$ are the desired eigenvalues. The eigenvectors are obtained by solving the corresponding systems of equations $AV(i) = \lambda(i)V(i)$. If the rows of the matrix are constant multiples of the first row, then all eigenvalues except one are zero and $\lambda(\max) = n$ where n is the size of the matrix. This occurs

Table 33.2. Comparison Scale

Intensity of Importance	Definition
1	Equal importance of both elements
3	Weak importance of one element over another
5	Essential or strong importance of one element over another
7	Demonstrated importance of one element over another
9	Absolute importance of one element over another
2, 4, 6, 8	Intermediate values between two adjacent importance values

in the ideal case. In practice, the weights cannot be given precisely and it is found that estimates of the weights by an expert will result in small errors. These small errors will result in small perturbations of the eigenvalues. The problem becomes one of finding the $\lambda(max)$ for this perturbed matrix. The eigenvectors will provide the proper relationship of the priorities of the pairwise comparison of the variables. The eigenvector measurement preserves ordinal preferences among the alternatives.

In the decision-making process, it is important to know how good the consistency is, and that it is not based on random selection. The consistency error is measured by its departure from n $(\lambda(max) \geqslant n)$.

The average, which is called the Consistency Index (CI), is defined as:

$$CI = \frac{\lambda(max)}{n-1} - n \qquad (33.2)$$

Some inconsistency is expected in the real world, based on human preferences. The overall consistency judgments are measured by comparison with a random matrix. In more precise terms, the resultant consistency index is compared with those of the same index obtained as an average over a large number of reciprocal matrices of the same order whose entries are random and range from 1 to 9 (see Table 33.3).

The Consistency Ratio (CR) can then be evaluated to determine how good the original expert estimations are by:

$$CR = CI/RI \qquad (33.3)$$

A consistency ratio of 10% or less is considered acceptable. If, it is more than 10%, the judgment may be more random than desired and possible should be revised.

This method is particularly applicable to safety-related analytical problems when limited data is available (Pizzo, 1985).

33.4 GAME THEORY

Operations research is a title given to quantitative methods that attempt to find solutions to decision problems governing the function or design of systems. Such systems are frequently systems of commerce or business. Nearly always, they involve interactions with humans either as operators or indirect controllers. The methods are not the traditional mathematical methods that are so common in the physical sciences. They may use these methods as well as a variety of statistical tools.

The publication of a book by J. von Neumann and O. Morgenstern in 1953, called the *Theory of Games and Economic Behavior* was the beginning of modern operations research. In this book, decision making was thought of as playing a game, either against another intelligent decision-making entity or against the en-

Table 33.3. Random Index (*RI*) of a Matrix

Size of Matrix	1	2	3	4	5	6	7	8	9	10
Random Consistency	.00	.00	.58	.90	1.12	1.24	1.32	1.41	1.45	1.49

vironment. From this original work has arisen a vast body of knowledge in decision making. We will examine a few of these methods that may be of use in safety decision-making.

When discussing decision-making as a game there are certain terms that are useful and should be defined. A *two-person game* is a game in which there are two decision-making entities. A *play* occurs when each of the decision-making entities makes a decision and an outcome results. A *payoff* is made and the decision-making entities are ready for the next decision. A payoff is the result of a decision by two competing decision-making entities. A *strategy* describes the sequence of decisions in a series of plays. A *zero-sum game* is a game in which the sum of losses and gains of all the decision-making entities is zero. A gambling game in which the proprietor takes a part of each play is not a zero-sum game.

It is customary and convenient to list the payoffs of a two-person, zero-sum game in a matrix. Since each party to the game may have a number of choices or alternatives upon which to decide for each play, a matrix provides a convenient format to display payoffs for the players.

We will examine several criterion for decision-making. *Minimax* or *maximin* are conservative criteria. These criteria place a conservative bound on possible losses or gains. *Expected value*, as a criterion, takes the traditional, most likely approach to optimizing decision strategies. An examination of a regret matrix will reveal an additional, modified, conservative criterion for decision-making.

33.5 PAYOFF MATRIX REDUCTION

An explanation of payoff matrix should begin with an examination of a payoff matrix that describes payoffs in a two-person, zero-sum game. Table 33.4 is such a matrix.

Orient yourself as A, playing against B. The signs in the matrix pertain to A. Thus, all plus signs are losses for B. Note that the payoffs in the matrix may be in dollars or other units of value such as measures of utility value. For instance, if A selected a_1 and B selects b_1, A will win perhaps \$500,000 or \$5 million. Such a payoff matrix presents a bewildering array of possible choices for A. However, it may be possible to simplify the matrix by the principle of dominance. An alternative

Table 33.4. Two-Person Payoff Matrix

		B			
		b_1	b_2	b_3	b_4
	a_1	5	4	5	6
	a_2	4	2	1	−7
A	a_3	2	−10	9	10
	a_4	5	4	3	2
	a_5	6	−2	7	3

Table 33.5. Reduced Payoff Matrix

		b_2	b_3	b_4
			B	
	a_1	4	5	6
A	a_3	-10	9	10
	a_5	-2	7	3

is dominated if each of its payoffs is less than the corresponding payoffs of another alternative. If an alternative is dominated it can be eliminated from the matrix. And, of course, if A observes an alternative is dominated, B will likewise eliminate it.

A then observes that alternative 2 is dominated by 1. Two is eliminated from the matrix. B observes that alternative 1 is dominated by 2. One is eliminated from the matrix. A then observes that alternative 4 is dominated by 1. Four is eliminated from the matrix. Table 33.5 is the reconstituted matrix.

B may then observe that alternative 2 dominates both 3 and 4. Three and four are eliminated leaving B's alternative 2. A will then observe that alternatives 3 and 4 are dominated by 1 and remove those. We have now reduced the original 5 by 4 payoff matrix to a 1 by 1. This is called a saddle point. It means that A will always play alternative 1 even knowing that B will always play 2. B will always play 2 even knowing that A will always play 1. For either party to do otherwise would be foolish. If B plays any other alternative than 2 while A is playing 1, B will lose more than 4. If A plays any other alternative than 1 while B is playing 2, A will win no more than 4 and perhaps less.

A saddle point will never be violated by playing minimax–maximin criteria. However, only a few payoff matrices have saddle points.

33.6   MINIMAX–MAXIMIN DECISION CRITERIA

To understand the minimax–maximin criteria of decision making we should examine another matrix. Table 33.6 is the new matrix.

The maximin strategy is played by A by examining each of his alternative payoffs for the minimum payoff. A notes this value for each alternative and selects the alternative that has the maximum of these minimums. If A does this for Table 33.6, alternative 4 will be selected. This assures A that no matter which alternative B plays, A will win no less than 2. If B plays the same conservative bounding strategy, but assuming the signs are reversed, B will play 3 assuring that B will lose no more than 7. Note that for both A and B these strategies may be unstable. If each notes the other playing the minimax-maximin strategy, they may be tempted to vary their selection in an effort to increase their payoffs or reduce their losses.

One may wonder why B plays a game in which losses will almost certainly occur. B might play because B has no choice but to play. In this sense we may

Table 33.6. Minimax–Maximin Payoff Matrix

		B			
		b_1	b_2	b_3	b_4
	a_1	9	3	5	−6
	a_2	4	2	1	3
A	a_3	2	−10	7	10
	a_4	5	4	3	2
	a_5	−6	8	−7	3

think of B as not an independent, intelligent, decision-making entity, but an environmental state. B may be the future sets of environmental states in which the decisions of A will pay off or not. If this is the case, A's purposes would be better served if the probabilities of B playing its choices could be estimated. By giving B free choice to play any of its alternatives, b_i, we are assuming that each of these choices is equal-likely. Table 33.7 is the same payoff matrix as Table 33.4, but in this case B is not free to chose and is constrained by the probabilities for each choice.

Given the probabilities of B's choices, A may calculate the $E(P)$ (expected payoff) of each of its choices.

$$E(P_{a1}) = .3(9) + .2(3) + .4(5) + .1(−6) = 4.7$$

We can see that the expected payoff of a_1 is greatest and thus A should chose a_1 if A is using expected value as the decision criterion. A's use of expected value as the criterion implies several things. Perhaps the most important consideration from A's point-of-view is that the occasional loss of 6 is acceptable. If such a loss is perhaps \$6E6 and would result in financial ruin for A, then A must reject expected value as the decision criterion, at least for this game. By playing expected value B has relatively high probabilities of winning 9 or 5 and a low probability of losing

Table 33.7. Minimax–Maximin Payoff Matrix with Probabilities

		B				
		b_1	b_2	b_3	b_4	
		.3	.2	.4	.1	$E(P)$
	a_1	9	3	5	−6	4.7
	a_2	4	2	1	3	2.3
A	a_3	2	−10	7	10	2.4
	a_4	5	4	3	2	3.7
	a_5	−6	8	−7	3	−2.7

6. A's playing of expected value should also imply that A will be playing repeatedly. In fact, if A can stand the loss of 6 and is playing repeatedly, A would be strongly advised to play expected value. Nevertheless, A's maximin criterion remains the preferred conservative criterion, assuring that A does no worse than win 2.

33.7 REGRET DECISION CRITERIA

Regret as a decision criterion implies that the decision maker must somehow find regrets and minimize them in decision making. Regrets are the loss a decision maker experiences in the decision made, compared to the payoff if the decision maker had known which environmental state would have prevailed and have made the optimum decision for this state. For instance, if we examine Table 33.6 we can see that if A selects a_3 and state b_1 occurs then A has a regret of 7. This is because if A had known that b_1 would occur a_1 could have been selected with a payoff of 9 rather than 2.

Following these principles for finding regrets, a new payoff table of regrets can be constructed from payoff Table 33.7. Table 33.8 illustrates.

We may now apply decision criteria to the regret matrix. The minimax regret criterion will result in selection of a_2. The minimax regret selection may be less conservative than maximin criterion on the original matrix.

Note that the minimum expected regret will result in selecting a_1. This was the same selection as expected value on the original payoff matrix. This will always be the case for the expected regret decision criterion. While expected regret does not provide a new decision, there is additional information that may be gleaned from this calculation. We see that the expected regret for alternative a_1 is 3.4. This is the upper bound on the amount you would be willing to pay for expert information as to which state will occur for each decision or play. We will call this value the *expected value of perfect information* (EVPI). There are methods that can be used to determine more precisely the amount you should pay for expert information about the future, given the record of the expert in making such predictions.

Table 33.8. Regret Payoff Matrix

				B		
		b_1	b_2	b_3	b_4	
		.3	.2	.4	.1	E(R)
	a_1	0	5	2	16	3.4
	a_2	5	6	6	7	5.8
A	a_3	7	18	0	0	4.3
	a_4	4	4	4	8	4.4
	a_5	15	0	14	7	9.3

33.8   MULTIATTRIBUTE DECISION MAKING

Referring back to the five questions that must be answered in decision making under uncertainty, we have addressed only question 5. We should briefly discuss how a payoff matrix is completed so the foregoing decision methods may be employed.

Imagine a decision maker faced with several alternatives from which to chose. Each alternative possesses a number of performance characteristics. These performance parameters describe how the system will perform. The performance parameters have multidimensions. If the payoff matrix is to be completed so that a decision can be made, these performance parameters must be evaluated and units of value placed on the many dimensions. Recall that the first four questions of the decision process required a definition of the magnitudes of the performance parameters and their variation under the states of the uncertain future. Having decided on the magnitudes of value of a system's performance parameters, we must sum them together for a particular alternative under a particular future state.

If we are considering a system in the development cycle the magnitudes of its performance parameters are not known. If the system is still a paper system there will be changes in these performance parameters as a result of future environmental states in which it is developed. Assume we are developing a new missile. Its performance parameters might be range, accuracy, and velocity, to name three. Early, and to a certain extent, throughout the development cycle, we will not know precisely the magnitudes of these performance parameters. Thus, if we are selecting a contractor for this missile we must estimate them under the several uncertain states that will obtain during the development of the missile.

Let us assume that we have a number of competing systems alternatives, a_i, and that these alternatives may be evaluated by a single quantity, u_{ik}, where k identifies a future state. This evaluator may vary with the future state S_k, and this future state will occur with probability P_k. The evaluator, u_{ik}, will contain evaluations of all the performance parameters that describe the system and upon which the decision should be made. Further they must contain the variation of the performance parameters and the variation of the values placed upon the performance parameters. Table 33.9 is such a general payoff matrix.

Table 33.9. System Payoff Matrix[a]

	P_1	P_2	P_3	P_4
	S_1	S_2	S_3	S_4
a_1	u_{11}	u_{12}	u_{13}	u_{14}
a_2	u_{21}	u_{22}	u_{23}	u_{24}
a_3	u_{31}	u_{32}	u_{33}	u_{34}
a_4	u_{41}	u_{42}	u_{43}	u_{44}

[a]*Note:* a_1 = alternative; u_{ik} = alternative evaluator; S_k = future environmental state; and P_k = environmental state probability.

Table 33.10. Performance Parameter Matrix

	S_1 P_1	S_2 P_2	S_k P_k
y_{i1}	$f(y_{i1})_1$	$f(y_{i1})_2$	$f(y_{i1})_k$
y_{i2}	$f(y_{i2})_2$	$f(y_{i2})_k$	
y_{ij}	$f(y_{ij})_1$	$f(y_{ij})_2$	$f(y_{ij})_k$

If Table 33.9 can be completed, we will be in a position to make a selection of the preferred alternative based upon the previously examined decision criteria. To complete Table 33.9 we will need two additional matrices. One will contain the magnitudes of the performance parameters for the system in question. The other will address the values we place on these performance parameters. Table 33.10 is the matrix that addresses the magnitudes of the performance parameters

where

$$i = \text{alternative}$$

$$j = \text{performance parameter}$$

$$k = \text{future state}$$

This one matrix represents the variation in the performance parameters for one alternative under each of the possible future states that may influence the magnitudes of the performance parameters. Note that each cell contains a function of the performance rather than a point value. Thus, the function, $f(y_{12})_3$, represents the variation of performance 2, of alternative 1, under future state 3. Such a function may take the shape of Figure 33.2.

If this performance parameter is range we can see what the most likely range will be or, of more importance, we can determine the minimum range with an associated probability.

A similar matrix to Table 33.10, must be developed for each alternative.

Next, we must place values on the performance parameters as they vary under the possible states of the uncertain future. This will require a matching matrix for each of the performance parameter matrices under each alternative.

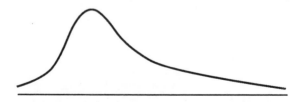

Figure 33.2   Performance parameter function $f(y_{12})_3$.

33.8.1   Utility Value Theory

Utility theory describes a body of knowledge by which a unidimensional measure of value can be placed on varied dimensions. These measures of value can be so sized that they consider our perception of the worth of costs or benefits.

Suppose that you are offered a bet in which you will flip a coin and lose $.75 if heads comes up and win $1.00 if it is tails. Undoubtedly you would make this bet. The expected payoff is positive ($.125). However, suppose that the bet involves losing $7500 and winning $10,000 on a single flip of a fair coin. Most individuals with limited resources would probably not take such a bet. The loss of $7500 would be too great to balance the winning of $10,000. If we are a large organization with proportionately large resources, we would be pleased to take such a bet and many more like it.

This simple example shows that the limited resource individual places disproportionately large negative value on losing $7500 and disproportionately small positive value on winning $10,000 or perhaps a combination of both. Furthermore, we can see that organizations will place different values on winnings and losses than individuals. We would expect to be able to draw a curve of the relationship between utility, as a measure of value, and the real dimension. Figure 33.3 is a conservative individual's utility curve of gambling for money.

We can see in Figure 33.3 that, as in the case of the bet, there is disproportionately small positive utility placed on winning and disproportionately large negative utility placed on losing. Note that there is a small positive utility in breaking even. An aggressive, optimistic gambler might have a utility curve similar to that of Figure 33.4.

Here we can see the disproportionately large positive utility in winning and vice versa in losing. Note the negative utility in breaking even. A large organization with large resources may have a utility curve similar to Figure 33.5.

In Figure 33.5, we can see that the organization is indifferent to winning and losing, valuing both proportionately.

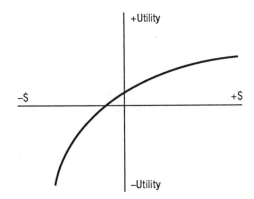

Figure 33.3 Conservative utility curve of gambling.

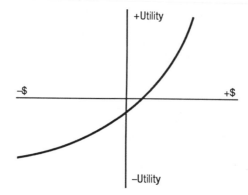

Figure 33.4 Optimistic gambler's curve.

Utility methods can also be applied to dimensions other than units of value ($). For instance, by asking appropriately structured questions of persons responsible for a missile's operational deployment, it is possible to find the utility curve for the range of a missile. Figure 33.6 is an example of such a curve.

We can see that the range of this missile has zero utility until it reaches some minimum range. At this range the utility rises quickly to the maximum needed range. After this range, the utility levels off or even declines.

33.8.2 Completing the Decision Matrices

We are concerned with the decision maker who wishes to make rational decisions involving large, complex systems during the concept or early design stages of system development. These decisions may involve selecting a preferred system safety bid or even selecting sets of safety countermeasures. Such a decision involves transforming multidimensional outcomes into a single utility measure. There is evidence to

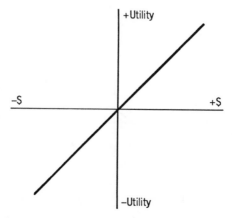

Figure 33.5 Large organization's utility curve.

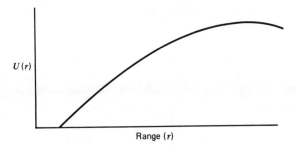

$U(r)$

Range (r)

Figure 33.6 Missile range utility curve.

suggest that the human mind has difficulty with such a problem, confusing the
utility values placed on dimensional distributions with personal bias. It is suggested
that quantifying human judgment with respect to expected outcomes of utility values
is a better procedure than using subjective judgment.

The values in the payoff matrix will become expected utility values of each
alternative. These expected values will combine all of the characteristics of performance
and utility, considering the uncertain future states that may prevail. Theoretically,
the utility values in the payoff matrix, each representing distributions of all performance
parameters and utility values, are created as in equation (33.4).

$$u_{ijk} = \int u(y_{ij})_k f(y_{ij})_k dy_{ij} \qquad (33.4)$$

The utility function $u(y_{ij})_k$, of equation (33.4) must be developed for each cell
of a utility matrix. Such a matrix must be created to match each of the performance
matrices for each alternative, a_i, as shown in Table 33.10. Table 33.11 is the outline
for such a matrix.

This method is most general and sufficiently flexible to cover many conditions.
For instance, if alternative future states of the environment do not influence either
the performance or utility distributions, then these columns in the corresponding
Tables will be identical.

*It is emphasized that both utility as a measure of value and probability as a
measure of uncertainty are inherent elements of the decision making. Uncertainty
and value judgments are present in any decision process whether or not they are
recognized. It would be well to make them explicit.*

Table 33.11. Utility Matrix

	P_1	P_2	P_3	P_k
	S_1	S_2	S_3	S_k
y_{i1}	$u(y_{i1})_1$	$u(y_{i1})_2$	$u(y_{i1})_3$	$u(y_{i1})_k$
y_{i2}	$u(y_{i2})_1$	$u(y_{i2})_2$	$u(y_{i2})_3$	$u(y_{i2})_k$
y_{i3}	$u(y_{i3})_1$	$u(y_{i3})_2$	$u(y_{i3})_3$	$u(y_{i3})_k$
y_{ij}	$u(y_{ij})_1$	$u(y_{ij})_2$	$u(y_{ij})_3$	$u(y_{ij})_k$

As systems become more complex and store more energy, their potential to do harm increases. This increasing potential for harm makes decisions in regard to the future safety of system operation ever more important. In recent years, it is fair to say that safety decisions carry the comparable weight, in the future financial success of an enterprise, to decisions concerning the operational features of the system. Thus, it would seem to be prudent to employ decision methods for safety that have been developed and used successfully in these other technical areas of system development.

Quite simply, the safety professional must move from making value judgments intuitively to the situation in which the elements of the decision process are exposed for management to examine and criticize. In this way, system safety can make a proper and successful contribution to the final success of a complex and costly endeavor.

EXERCISES FOR PART VI

1. Describe a safety situation that would be properly addressed by decision-making methods.

2. Create a decision-making method that borrows from the Delphi method but is significantly different.

3. Draw a utility curve for a flight of a passenger-carrying aircraft as the number of passengers increases toward capacity on the positive and of an accident in which all are killed on the negative side.

4. An evaluation is made of contractors system safety proposals for a major new system. The proposals are ranked under four possible future states on a utility scale from negative 10 to positive 10. The probabilities of the future states are $P_1 = .1$, $P_2 = .3$, $P_3 = .2$, and $P_4 = .4$. These are the evaluations of the contractors.

Contractor 1 has a utility of 10 under state 1, 6 under 2, 2 under 3, and −4 under 4.

Contractor 2 has a utility of 2 under state 1, 7 under 2, −1 under 3, and 10 under 4.

Contractor 3 has a utility of 1 under state 1, 4 under 2, 2 under 3, and 0 under 4.

Contractor 4 has a utility of −8 under state 1, 10 under 2, −2 under 3, and 8 under 4.

Contractor 5 has a utility of −1 under state 1, 9 under 2, 8 under 3, and 9 under 4.

(a) Select the preferred contractor by the criterion of maximin.

(b) Select the preferred contractor by the criterion of expected value.

(c) Select the preferred contractor by the criterion of minimax regret.

(d) What is the expected value of perfect information?

(e) What is the loss of expected value if maximin is used as the selection criterion rather than expected value?

(f) What is the loss of expected value if maximin is used as the selection criterion rather than minimax regret?

REFERENCES AND BIBLIOGRAPHY

Ben-Horim, M. and Levy, H., *Statistics, Decision and Applications in Business and Economics*. Random House, New York, 1984.

Brown, R. V., Kahr, A. S., and Peterson, C., *Decision Analysis for the Manager*. Holt, Rinehart and Winston, New York, 1974.

Gilbreath, G. H. and Van Marte, J. G., *Statistics for Business Economics*. Business Publications, Dallas, TX, 1980.

Gordon, G. and Pressman, I., *Quantitative Decision-Making for Business*. Prentice-Hall Inc., Englewood Cliffs, NJ, 1978.

Krone, R. M., *Systems Analysis and Policy Sciences*. Wiley, New York, 1980.

Levy, H. and Sarnat, M., *Investment and Portfolio Analysis*. Wiley, New York, 1972.

Neuman, J. von and Morgenstern, O., *Theory of Games and Economic Behavior*, rev. ed.. Princeton University Press, Princeton, NJ, 1953.

Pizzo, J. T. and Moriarity, B. M., "Quantification of Subjective Judgments for Safety Estimates. Use of Metric Matrix Hierarchical Analysis (MMHA). *Proceedings of System Safety Society Conference*, 1985.

Winkler, R. L. and Hays, W. L., *Statistics: Probability Inference and Decision*, 2nd ed. Holt, Rinehart and Winston, New York, 1975.

APPENDIX A

STATISTICAL TABLES

Table A.1. Areas under the normal curve

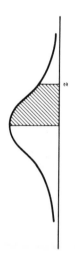

z	0.00	0.01	0.02	0.03	0.04	0.05	0.06	0.07	0.08	0.09
0.0	0.00000	0.00399	0.00798	0.01197	0.01596	0.01995	0.02393	0.02791	0.03189	0.03587
0.1	0.03984	0.04380	0.04777	0.05173	0.05568	0.05963	0.06357	0.06750	0.07143	0.07535
0.2	0.07927	0.08317	0.08707	0.09096	0.09484	0.09871	0.10257	0.10642	0.11026	0.11409
0.3	0.11791	0.12172	0.12551	0.12930	0.13307	0.13682	0.14057	0.14430	0.14802	0.15172
0.4	0.15541	0.15909	0.16275	0.16639	0.17002	0.17363	0.17723	0.18081	0.18437	0.18792
0.5	0.19145	0.19496	0.19846	0.20193	0.20539	0.20883	0.21225	0.21565	0.21903	0.22239
0.6	0.22574	0.22906	0.23236	0.23564	0.23890	0.24214	0.24536	0.24856	0.25174	0.25489
0.7	0.25803	0.26114	0.26423	0.26730	0.27034	0.27337	0.27637	0.27935	0.28230	0.28523
0.8	0.28814	0.29103	0.29389	0.29673	0.29955	0.30234	0.30511	0.30785	0.31057	0.31327
0.9	0.31594	0.31859	0.32123	0.32382	0.32640	0.32895	0.33148	0.33398	0.33646	0.33892
1.0	0.34135	0.34375	0.34614	0.34850	0.35084	0.35315	0.35544	0.35770	0.35994	0.36215
1.1	0.36434	0.36651	0.36865	0.37077	0.37287	0.37494	0.37699	0.37901	0.38101	0.38299
1.2	0.38494	0.39697	0.38878	0.39066	0.39252	0.39436	0.39618	0.39797	0.39974	0.40148
1.3	0.40321	0.40491	0.40659	0.40825	0.40989	0.41150	0.41309	0.41466	0.41621	0.41774
1.4	0.41925	0.42074	0.42220	0.42365	0.42507	0.42648	0.42786	0.42923	0.43057	0.43189

z										
1.5	0.43320	0.43448	0.43575	0.43700	0.43822	0.43943	0.44062	0.44180	0.44295	0.44408
1.6	0.44520	0.44632	0.44739	0.44845	0.44950	0.45053	0.45154	0.45254	0.45352	0.45449
1.7	0.45543	0.45637	0.45728	0.45818	0.45907	0.45994	0.46079	0.46163	0.46246	0.46327
1.8	0.46407	0.46485	0.46562	0.46637	0.46711	0.46784	0.46855	0.46925	0.46994	0.47061
1.9	0.47128	0.47193	0.47256	0.47319	0.47380	0.47440	0.47499	0.47557	0.47614	0.47670
2.0	0.47724	0.47778	0.47830	0.47881	0.47932	0.47981	0.48029	0.48076	0.48123	0.48168
2.1	0.48213	0.48256	0.48299	0.48340	0.48381	0.48421	0.48460	0.48499	0.48536	0.48573
2.2	0.48609	0.48644	0.48678	0.48712	0.48744	0.48776	0.48808	0.48839	0.48869	0.48898
2.3	0.48926	0.48954	0.48982	0.49009	0.49035	0.49060	0.49085	0.49110	0.49133	0.49158
2.4	0.49179	0.49201	0.49223	0.49244	0.49265	0.49285	0.49304	0.49323	0.49342	0.49360
2.5	0.49378	0.49395	0.49412	0.49429	0.49445	0.49460	0.49476	0.49491	0.49505	0.49519
2.6	0.49533	0.49546	0.49559	0.49572	0.49585	0.49597	0.49608	0.49620	0.49631	0.49642
2.7	0.49652	0.49663	0.49673	0.49683	0.49692	0.49701	0.49710	0.49719	0.49728	0.49736
2.8	0.49744	0.49752	0.49759	0.49767	0.49774	0.49781	0.49788	0.49794	0.49801	0.49807
2.9	0.49813	0.49819	0.49824	0.49832	0.49835	0.49841	0.49846	0.49851	0.49856	0.49860
3.0	0.49865	0.49869	0.49873	0.49877	0.49881	0.49885	0.49889	0.49893	0.49896	0.49900
3.1	0.49902	0.49906	0.49909	0.49912	0.49915	0.49918	0.49921	0.49923	0.49926	0.49929
3.2	0.49931	0.49933	0.49936	0.49938	0.49940	0.49942	0.49944	0.49946	0.49948	0.49950
3.3	0.49951	0.49953	0.49955	0.49956	0.49958	0.49959	0.49961	0.49962	0.49964	0.49965
3.4	0.49966	0.49967	0.49969	0.49970	0.49971	0.49972	0.49973	0.49974	0.49975	0.49976
3.5	0.49977	0.49977	0.49978	0.49979	0.49980	0.49981	0.49981	0.49982	0.49983	0.49983
3.6	0.49984	0.49985	0.49985	0.49986	0.49986	0.49987	0.49987	0.49988	0.49988	0.49989
3.7	0.49989	0.49990	0.49990	0.49990	0.49991	0.49991	0.49991	0.49992	0.49992	0.49992
3.8	0.49993	0.49993	0.49993	0.49994	0.49994	0.49994	0.49994	0.49995	0.49995	0.49995
3.9	0.49995	0.49995	0.49996	0.49996	0.49996	0.49996	0.49996	0.49996	0.49997	0.49997

Table A.2. Fractiles of the χ^2 distribution

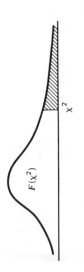

ν	$1 - F(\chi^2)$								
	0.75	0.50	0.25	0.20	0.15	0.10	0.05	0.025	0.01
1	0.1015	0.4549	1.323	1.642	2.072	2.706	3.841	5.024	6.635
2	0.5754	1.386	2.773	3.219	3.794	4.605	5.991	7.378	9.210
3	1.213	2.366	4.108	4.642	5.317	6.251	7.815	9.348	11.345
4	1.923	3.357	4.385	5.989	6.745	7.779	9.488	11.143	13.277
5	2.675	4.351	6.626	7.289	8.115	9.236	11.071	12.833	15.086
6	3.455	5.381	7.841	8.558	9.446	10.645	12.592	14.449	16.812
7	4.255	6.346	9.037	9.803	10.746	12.017	14.067	16.013	18.475
8	5.071	7.344	10.219	11.030	12.027	13.362	15.507	17.535	20.090
9	5.899	8.343	11.389	12.242	13.288	14.684	16.919	19.023	21.666
10	6.737	9.342	12.549	13.442	14.534	15.987	18.307	20.483	23.209
11	7.584	10.341	13.701	14.631	15.767	17.275	19.675	21.920	24.725
12	8.438	11.340	14.845	15.812	16.989	18.549	21.026	23.337	26.217
13	9.299	12.340	15.984	16.985	18.202	19.812	22.362	24.736	27.688
14	10.165	13.339	17.117	18.151	19.406	21.064	23.685	26.119	29.141
15	11.037	14.339	18.245	19.311	20.603	22.307	24.996	27.488	30.578

	z_α								
	−0.674	0.000	0.674	0.841	1.036	1.282	1.645	1.960	2.327
16	11.912	15.339	19.369	20.465	21.793	23.542	26.296	28.845	32.000
17	12.792	16.338	20.489	21.615	22.977	24.769	27.587	30.191	33.409
18	13.675	17.338	21.605	22.760	24.155	25.989	28.869	31.526	34.805
19	14.562	18.338	22.718	23.900	25.329	27.204	30.144	32.852	36.191
20	15.452	19.337	23.828	25.038	26.498	24.412	31.410	34.170	37.566
21	16.344	20.337	24.934	26.171	27.662	29.615	32.671	35.479	38.932
22	17.240	21.337	26.039	27.301	28.822	30.813	33.924	36.781	40.289
23	18.137	22.337	27.141	28.429	29.979	32.007	35.173	38.076	41.638
24	19.037	23.337	28.241	29.553	31.132	33.196	36.415	39.364	42.980
25	19.939	24.337	29.339	30.675	32.282	34.382	37.653	40.647	44.314
26	20.843	25.336	30.435	31.795	33.429	35.563	38.885	41.923	45.642
27	21.749	26.336	31.528	32.912	34.574	36.741	40.113	43.194	46.963
28	22.657	27.336	32.621	34.027	35.715	37.916	41.337	44.461	48.278
29	25.567	28.336	33.711	35.139	36.854	39.088	42.557	45.722	49.588
30	24.478	29.336	34.800	36.250	37.990	40.256	43.773	46.979	50.892
z_α	−0.674	0.000	0.674	0.841	1.036	1.282	1.645	1.960	2.327

Note: For degrees of freedom greater than 30 use the expression:

$$(z_\alpha + \sqrt{2\nu - 1})^2$$

APPENDIX B

ANSWERS TO QUANTITATIVE EXERCISES

PART II, STATISTICAL METHODS

Probability

1. (a) 7/120, (b) 7/40, (c) 11/60, (d) 2/7

2. (a) 63/1000, (b) 189/1000, (c) 27/125, (d) 9/50

3. (a) 28/55, (b) 12/55, (c) 13/55, (d) 41/55, (e) 32/33, (f) .9950, (g) 1.00

4. (a) 1/4, (b) 7/16, (c) 5/8, (d) 37/64, (e) 5/8

5. (a) 2/5, (b) 1.00, (c) 7/10, 3/10, (d) 98/125

Statistical Measures

1. 4.378, 3.509, 5.8, 5.8

2. 2.281, 2.150

3. $1.305E-2$, $5.293E-5$

Statistical Distributions

Binomial

1. .513, .357, .109, $1.890E-2$, .130

2. (a) $2.622E-2$, (b) .930

3. $8.619E-2$

4. $7.071E-2$

5. (a) $4.8E-2$, (b) $6.691E-3$, (c) $2.304E-3$

6. (a) $.01$, (b) $4.901E-2$, (c) $9.802E-4$, (d) $3.915E-3$

Normal

1. (a) $3.150E-2$, (b) $2.200E-3$, (c) $.966$, (d) 712.3

2. (a) $.817$, (b) 753.6

3. (a) $5.480E-2$, (b) 14

4. 770

Poisson

1. (a) $.905$, (b) $9.516E-2$, (c) $3.351E-3$

2. (a) $.301$, (b) $.513$, (c) $1.015E-5$

3. (a) 30, (b) 30

Chi-square Confidence Bounding

1. (a) $7.847E-4$, (b) $3.388E-4$, (c) $.524$

2. (a) $8.871E-4$, (b) 1, (c) $3.838E-4$

3. (a) $1.336E-2$, (b) 108, (c) $.80$

4. (a) $\$78.27E6$, (b) $>.99$, (c) $2.942E-6$, (d) $\$44.41E6$

PART III, NETWORK ANALYSIS

3. (a) $.2731$, (b) $.6327$

4. $3.205E-2$

5. $2.322E-3$

6. (a) $.1$
   (b) 0.0
   (c) $.2$
   (d) $.6$
   (e) $.69$
   (f) $.40$
   (g) $.3$
   (h) $.9$

(i) .1
(j) .9
(k) .2
(l) .2
(m) 0.0
(n) .5
(o) .1
(p) 0.0

7. (a) .995, .99998
   (b) .5%

8. .2793

9. $4.979E-2$

10. $4.875E-3$

PART IV, HAZARD ANALYSIS

8. (a) *CD*, *AE*, *AC*
   (b) *ABCD*, *ABEF*, *BFGH*
   (c) *ABD*, *ACEFGH*

9 (a) *AB*, *ADE*, *ADF*, *ADG*
  (b) *B*, *AF*, *AEG*, *AEH*, *DF*, *DEG*, *DEH*
  (c) *ACE*, *ACF*, *ACG*, *ACH*, *ADE*

12. (a) Event—$1.148E-3$, $5.937E-4$, $2.514E-3$, $1.594E-2$, $5.371E-3$
       Cut set—$6.819E-7$, $4.603E-8$, $5.083E-8$
       Top—$7.788E-7$
    (b) Event—.284, $7.531E-2$, .407, .999, .479
       Top—.173
    (c) Event—.284, $7.531E-2$, .407, .999, .479
       Top—$3.702E-4$
    (d) $I_{\bar{a}}$—.935, .941, $5.911E-2$, .124, $6.527E-2$
       $I_{\bar{r}}$—.899, .878, .122, .223, .101

PART V, RISK ANALYSIS

7. $699

8. $152, Yes

9. $-\$4.345E5$, No

10. $-\$4.23E4$, No

11. $\$1.98E4$, Yes

PART VI, DECISION THEORY

4. (a) a_3, (b) a_5, (c) a_2, (d) 1.7, (e) 5.2, (f) 1.9

INDEX